Advance pr

# Intuition in Mathematics and Physics

Desmet understands in a profound way that modern science has been so formulated as to conflict with basic human intuitions. As a result, overall, education has had to socialize students into articulating as scientific truth ideas that are strictly speaking not believable. These essays make clear that Whitehead rejected this whole approach and undertook to show that and how evidence uncovered by science can be explained in ways that are coherent with inescapable human intuitions. When Whitehead's program is understood in this way it can be seen as a gift of immeasurable value. This book goes far to demonstrate what the gift is and the possibility of accepting it.

**John B. Cobb, Jr.**
Claremont School of Theology
Author, *Spiritual Bankruptcy*

Ronny Desmet has put together a truly wonderful anthology. In his introduction and other essays — which are more than worth the price of the book — he insightfully explains the overall importance of Whitehead's philosophical vision: Showing how mathematics and the physical sciences can and should be developed in ways that reflect the "deep intuitions of humankind," which we inevitably presuppose, rather than being antithetical to them. The other essays demonstrate various ways in which Whitehead's philosophy, by developing concrete experiences rather than juggling with abstractions, can overcome the damage caused by the modern scientific worldview.

**David Ray Griffin**
Claremont School of Theology
Author, *God Exists but Gawd Does Not*

42

This intriguing volume convenes nine scientists, mathematicians, and philosophers who collectively dissent from the reductivist and atemporal worldview offered as the sole proper inference from advances in contemporary science. They find that prevailing worldview counter-intuitive, incoherent, and utterly incapable of adequately linking up the specific insights of the physical and biological sciences with the forms of mathematical reasoning used to model behavior in these interconnected domains. They refuse to acquiesce in pedagogical efforts to "tame" intuition (involving what the philosopher Whitehead termed the "pathological demotion" of the deep intuitions of humankind, as if these were naught but illusions). Their essays instead champion notions of genuine physical and biological development and transformation, creative novelty, and aesthetic criteria of "elegance" and "beauty" as integral components of (rather than merely epiphenomenal "psychic additions" to) the natural world, evident in otherwise-diverse phenomena ranging from quantum mechanical "entanglement" and information theory to neurobiology and the non-reductionist biomathematics (or "integral biomathics") of Simeonov. These are truly fascinating explorations of what, once again, the philosopher Whitehead termed "Nature Alive."

**Dr. George R. Lucas, Jr.**
Reilly Center for Science, Technology & Value
Notre Dame University
Author *Ethics & Cyberwarfare* (Oxford University Press, Fall 2016)

In the late 19th and early 20th centuries something of a crisis arose in mathematics as intuition was replaced by logic as the foundation of this discipline. Whitehead was a major figure in this transition, but he did not jettison intuition. This collection of well written and instructive essays helps us to find the proper place for intuition not only in mathematics, but in all of the fields in which Whitehead was interested. Desmet is to be thanked for putting together this valuable collection of essays.

**Dan Dombrowski**
Seattle University
Author, *A History of the Concept of God*

*Intuition in Mathematics and Physics* is a collection of essays inspired by Alfred North Whitehead's critique of scientific materialism and his philosophy based on the idea that nature is a process of becoming alive with feeling. This international group of scientists, mathematicians and philosophers revitalize the fascinating implications of this creative intuition for specialized topics in physics, neurobiology and mathematics by focusing attention on the primacy of concrete, lived experience as a foundation for theoretical constructions. This perspective is a persuasive alternative to the lingering effects of positivism and reductionism in prevailing orthodoxy and holds out a promise for an ecological civilization.

**Leemon B. McHenry,**
California State University, Northridge
Author, *The Event Universe*

Dear Diana, Rita & Ron,

A book within which
I have an essay.

Dilip Hiro

# Intuition in Mathematics & Physics

# Intuition in
# Mathematics &
# Physics

## A Whiteheadian Approach

RONNY DESMET

EDITOR

PROCESS CENTURY PRESS

ANOKA, MINNESOTA 2016

*Intuition in Mathematics and Physics: A Whiteheadian Approach*

Process Century Press
RiverHouse LLC
802 River Lane
Anoka, MN 55303

Process Century Press books are published in association with the International Process Network.

Cover design: Susanna Mennicke

Toward Ecological Civilization Series, Volume X
Jeanyne B. Slettom, Series Editor

ISBN 978-1-940447-13-1
Printed in the United States of America

We live in the ending of an age. But the ending of the modern period differs from the ending of previous periods, such as the classical or the medieval. The amazing achievements of modernity make it possible, even likely, that its end will also be the end of civilization, of many species, or even of the human species. At the same time, we are living in an age of new beginnings that give promise of an ecological civilization. Its emergence is marked by a growing sense of urgency and deepening awareness that the changes must go to the roots of what has led to the current threat of catastrophe.

In June 2015, the 10th Whitehead International Conference was held in Claremont, CA. Called "Seizing an Alternative: Toward an Ecological Civilization," it claimed an organic, relational, integrated, nondual, and processive conceptuality is needed, and that Alfred North Whitehead provides this in a remarkably comprehensive and rigorous way. We proposed that he could be "the philosopher of ecological civilization." With the help of those who have come to an ecological vision in other ways, the conference explored this Whiteheadian alternative, showing how it can provide the shared vision so urgently needed.

The judgment underlying this effort is that contemporary research and scholarship is still enthralled by the 17th century view of nature articulated by Descartes and reinforced by Kant. Without freeing our minds of this objectifying and reductive understanding of the world, we are not likely to direct our actions wisely in response to the crisis to which this tradition has led us. Given the ambitious goal of replacing now dominant patterns of thought with one that would redirect us toward ecological civilization, clearly more is needed than a single conference. Fortunately, a larger platform is developing that includes the conference and looks beyond it. It is named Pando Populus (pandopopulous.com) in honor of the world's largest and oldest organism, an aspen grove.

As a continuation of the conference, and in support of the larger initiative of Pando Populus, we are publishing this series, appropriately named "Toward Ecological Civilization."

-John B. Cobb, Jr.

# Table of Contents

# Abbreviations of Whitehead's Writings

AE   *The Aims of Education and Other Essays.* New York: The Free Press, 1929/1967.

AI   *Adventures of Ideas.* New York: The Free Press, 1933/1967.

ADG   *The Axioms of Descriptive Geometry.* New York: Dover Publications, 1907/2005.

APG   *The Axioms of Projective Geometry.* Cambridge at the University Press, 1906/1913.

CN   *The Concept of Nature.* Cambridge: Cambridge University Press, 1920/1986.

ESP   *Essays in Science and Philosophy.* Westport, Connecticut: Greenwood Press, 1947/1968.

FR   *The Function of Reason.* Boston: Beacon Press, 1929/1958.

IM   *Introduction to Mathematics.* Oxford: Oxford University Press, 1911/1958.

MT   *Modes of Thought.* New York: The Free Press, 1938/1968.

NL   *Nature and Life.* Cambridge: Cambridge University Press, 1934/2011.

OT   *The Organisation of Thought: Educational and Scientific* Westport, Connecticut: Greenwood Press, 1917/1974.

PM   *Principia Mathematica to *56.* Co-authored with Bertrand Russell. Cambridge at the University Press, 1910/1970.

PNK   *An Enquiry Concerning the Principles of Natural Knowledge.* New York: Dover Publications, 1919/1982.

PR   *Process and Reality.* Corrected Edition. Edited by D. R. Griffin and D. W. Sherburne. New York: The Free Press, 1929/1985.

R        *The Principle of Relativity with Applications to Physical Science*
         New York: Dover Publications, 1922/2004.

RM       *Religion in the Making.* New York: Fordham University Press,
         1926/1996.

S        *Symbolism: Its Meaning and Effect.* New York: Fordham
         University Press, 1927/1985.

SMW      *Science and the Modern World.* New York: The Free Press,
         1925/1967.

UA       *A Treatise on Universal Algebra with Applications.* New York:
         Hafner Publishing Company, 1898/1960.

# Introduction

RONNY DESMET

This book is an anthology of ten essays written in the afterglow of the workshop on intuition in mathematics and physics at the 10<sup>th</sup> International Whitehead Conference in Claremont, California (June 2015). The common aim of its authors is to promote Whitehead's philosophical endeavor to bridge the modern gap between the abstractions of science and the deep intuitions of humankind. This gap is one of the major causes of the human alienation from nature and of the ensuing ecological crisis. The endeavor of bridging it is an important step toward an ecological society. Consequently, the limited aim of this anthology is part of the more encompassing ecological concern that motivated the conference.

In the first part of the introduction, I sketch the genesis of the modern gap between science and intuition as well as the mode of philosophizing that Whitehead developed to prevent the bifurcation of reality into the world of science and that of intuition. In part one, I mainly rely on Whitehead's *Science and the Modern World*, which explains its title.

In the second part of the introduction, I give an outline of how Whitehead's way of doing philosophy has led, in the first half of the 20<sup>th</sup> century, to the emergence of a scientific as well as intuitive cosmology—Whitehead's process cosmology. In part two, I mainly rely on Whitehead's *Nature and Life*, which again explains the corresponding title.

In the third part of the introduction, which has the same title as this anthology as a whole, I introduce the ten chapters that follow. This introduction reveals the diversity of the ten essays, but also makes clear that they do indeed have a common aim; namely, to highlight developments in 20th- and early 21st-century science and philosophy that have the potential to support, or even further, Whitehead's philosophical integration of the abstractions of mathematics and physics with the deep intuitions of humankind.

## SCIENCE AND THE MODERN WORLD

Seventeenth-century mathematicians and physicists excluded from the scope of their research our deep intuitions that nature is alive, creative, organic, purposive, and valuable, in order to focus on nature's lifeless manifestations, its recurrent behavior patterns, its mechanical laws, its efficient causality, and its sheer factuality. As a method of research this exclusion is justifiable, provided we recognize the limitations involved. However, with the overwhelming success of mathematics and physics, the limitations of their method have faded into oblivion, and their methodological exclusion has tacitly been transformed into an ontological exclusion. In other words, the method of abstraction in mathematics and physics has been so fruitful that it has led many of its practitioners and supporters to deny the existence in nature of the very things of which abstraction is made: vivid experience, creative novelty, organic relatedness, final causality, and aesthetic and moral values. This has led to a scientific worldview in which our intuition of feeling in nature must be taken as a sign that we have not yet completely overcome the anthropomorphic fallacy; our intuition of unforeseen novelty in nature as a sign of our lack of predictive power; our intuition of the internal relatedness of nature as a sign of our lack of knowledge of the external interplay of essentially isolated bits of matter in motion; our intuition of purposive behavior in nature as a measure of our ignorance with respect to the efficient causes that jointly determine all behavior in nature; and our intuitions of beauty and goodness in nature as purely subjective reactions to objective phenomena which are neither beautiful nor ugly, and neither good nor bad.

In *Science and the Modern World*, Whitehead used the term "cosmology" instead of "worldview." According to him, a particular scientific cosmology has been dominant since the seventeenth century, namely:

> the fixed scientific cosmology which presupposes the ultimate fact of an irreducible brute matter, or material, spread through space in a flux of configurations. In itself such a material is senseless, valueless, purposeless. It just does what it does do, following a fixed routine imposed by external relations which do not spring from the nature of its being. (SMW 17)

Whitehead called this dominant cosmology, "scientific materialism."

In fact, the modern ontological exclusion is more encompassing than just sketched because the successful and inspiring theories of mathematicians and physicists also make abstraction from aspects of nature that their practices—their very method and even their initial motivation—presuppose.

In practice, the empirical method of verification or falsification by observation, no matter how high-tech, relies on sense data. In theory, however, mathematical physicists abstract from most of the data of our five senses (sight, hearing, smell, taste, and touch) to focus on the colorless, soundless, odorless, and tasteless structural aspects of nature. Consequently, in a philosophical worldview inspired not by the actual practices of mathematical physicists, but merely by their theoretical results, nature—already stripped from its "tertiary" qualities (aesthetical, ethical, and religious values)—is further reduced to the scientific world of "primary" qualities (mathematical quantities and relationships such as the amplitude, length, and frequency of mathematical waves and their relatedness), and this scientific world is bifurcated from the intuitive world of "secondary" qualities (colors, sounds, etc.). Moreover, the scientific world is supposed, ultimately, to fully explain the intuitive world (so that, for example, colors end up as being nothing more than electromagnetic waves of certain frequencies). Whitehead spoke of "the bifurcation of nature into two systems of reality" (CN 30) to denote the strategy—originating with Galileo, Descartes, Boyle and Locke—of bifurcating nature into the essential reality of primary qualities and the non-essential reality of "psychic additions" or secondary qualities, ultimately to be explained away in terms of primary qualities. Whitehead sided with Berkeley in arguing that the primary/secondary distinction is not tenable (CN 43–44), that all qualities are "in the same boat, to sink or swim together" (CN 148), and that, for example, "the red glow of the sunset should be as much part of nature as are the molecules and electric waves by which men of science would explain the phenomenon" (CN 29).[1] Contrary to Berkeley and Whitehead, for promoters of the bifurcation of nature, our intuition of the colorfulness of nature, for example, must be taken as a sign of "the byplay of the perceiving mind" (CN 30), which clouds a truly scientific worldview.

Whitehead described the philosophical outcome of the bifurcation of nature as follows:

> The primary qualities are the essential qualities of substances whose spatio-temporal relationships constitute nature. . . . The occurrences of nature are in some way apprehended by minds . . . But the mind in apprehending also experiences sensations which, properly speaking, are qualities of the mind alone. These sensations are projected by the mind so as to clothe appropriate bodies in external nature. Thus the bodies are perceived as with qualities which in reality do not belong to them, qualities which in fact are purely the offspring of the mind. Thus nature gets credit which should in truth be reserved for ourselves: the rose for its scent: the nightingale for his song: and the sun for his radiance. The poets are entirely mistaken. They should address their lyrics to themselves, and should turn them into odes of self-congratulation on the excellency of the human mind. Nature is a dull affair, soundless, scentless, colourless; merely the hurrying of material, endlessly, meaninglessly. (SMW 54)

On top of that, whereas in practice what really motivates people to become mathematical physicists is the search for causal explanations, in theory, they often adhere to Hume's analysis of causality. Because there are *no sensory observations of* cause-and-effect happenings in nature, Hume holds that there are *no* such happenings at all. Causation is just a habit of thought, mysteriously (non-causally) emerging from the sensory observations of recurring successions of events. Kant's theory of causality is of the same family as Hume's. According to Kant, causality is just one of the categories of understanding applied to the chaos of sense data — data mysteriously (non-causally) emerging from an unknowable world *in itself* — to turn that chaos into a knowable world *for us*. Whether causation is conceived as "a habit of thought," or as "a category of thought," in both cases it is taken as revealing not the intrinsic relatedness of nature, but some additive operation of the human psyche (s 39–40). Hume's theory of causality seemed to be ratified by Newton's theory of gravity, which offered no causal explanation, but only a mathematical description of gravitational phenomena. And what Newton considered to be a shortcoming of his theory (the lack of causal explanation) gradually became a legitimation (especially after the rise of quantum theory) to start saying that the aim of mathematical physics should be restricted to giving mathematical descriptions of nature, while forgetting its original explanatory aim. Whitehead sided with Einstein in not wanting to give up the search for causal explanations. Whitehead refused to follow the majority of scientific philosophers who further reduce nature to a scientific world of causally unrelated bits of matter in motion, and who bifurcate this scientific world from the intuitive world of allegedly thought-imposed causality. Contrary to Einstein and Whitehead, for the heirs of Hume, our intuition of the causal relatedness of nature must be taken as a sign of our metaphysical urge to impose intrinsic relatedness and meaning where actually none exists — yet another demonstration of our intuitive resistance to a truly scientific worldview.

In some scientists "the instinctive conviction in the existence of . . . an *Order of Nature*" incoherently coexists with the philosophical claim that Hume was right to deny the causality of nature — Whitehead highlighted "this strange contradiction in scientific thought," and wrote:

> Since the time of Hume, the fashionable scientific philosophy has been such as to deny the rationality of science. This conclusion lies upon the surface of Hume's philosophy. Take, for example, the following passage from Section IV of his *Inquiry Concerning Human Understanding*:

In a word, then, every effect is a distinct event from its cause. It could not, therefore, be discovered in the cause; and the first invention or conception of it, *a priori*, must be entirely arbitrary.

If the cause in itself discloses no information as to the effect, so that the first invention of it must be *entirely* arbitrary, it follows at once that science is impossible, except in the sense of establishing *entirely arbitrary* connections, which are not warranted by anything intrinsic to the natures either of causes or effects. Some variant of Hume's philosophy has generally prevailed among men of science. (SMW 4)

We have many deep intuitions of nature: our intuitions of nature being alive; of its creativity and novelty; of its internal relatedness and its efficient as well as final causality; and of its secondary and tertiary qualities, ranging from blue to beauty, from green to grace, and from magenta to majesty. Sweeping aside, again and again, all these intuitions as mere signs of the imperfection of our current scientific understanding; presenting, again and again, the history of science as a succession of victories over the army of our misleading intuitions; has amounted to the widely shared (modern, Western) belief that nature is a dead and dull affair, fragmented and meaningless, colorless and valueless. Moreover, it has also resulted in our alienation from nature. How could we, holding fast to our own status as alive, creative, relational, causally determined, self-determined, colorful, and valuable beings, ever feel at home in a natural environment that is none of these?

In an attempt to overcome the resulting alienation from nature, we can claim, and many philosophers have done so, that all the intuitions we mentioned are illusions not only with respect to nonhuman nature, but also with respect to human beings. Moreover, such extension of the scope of scientific materialism from all *nonhuman* creatures of nature to *all* of them, including human beings, is coherent with the rise and acceptance of the theory of evolution, and with the associated awareness that humans are indeed fully part of nature. But it is hard not to hold fast to the human status as described above and to put aside our deepest intuitions of ourselves as nothing but illusions. In fact, it has proven impossible to consistently deny what we cannot help believing, and sustained attempts to do so anyway have resulted in self-alienation.

Consequently, even more philosophers have tried to overcome this disenchanting and reductionist monism. Some of them promoted an alternative monism in which the human mind is not part of nature, but nature is somehow part of an all-encompassing mind. But this "somehow" has proven to be "too much divorced from the scientific outlook" (SMW 63). So the majority of philosophers trying to overcome disenchantment and reductionism have tried to escape both the scientific but counter-intuitive material monism, and the slightly more intuitive but less scientific mental monism, by accepting and promoting some magical, or ontological, or epistemological material/mental dualism. According to these thinkers, despite the theory of evolution, we cannot be fully part of the natural world revealed by mathematics and physics. The human psyche or mind or soul must be part of some non-natural world: the supernatural world as opposed to the natural world, a magical escape; or the world of mind as opposed to the world of matter, an ontological escape that can be linked to Descartes; or the practical world of values as opposed to the theoretical world of facts, an epistemological escape that can be associated with Hume, Kant, or Wittgenstein. But the comfort of being able to switch back and forth from a theoretical to a practical perspective on reality, and from a value-free to an evaluative perspective, has not proven to be an enduring consolation. Dualism cannot satisfy our urge for coherence.

It seems as if we are stuck with only three options — material monism, mental monism, and some form of material/mental dualism — all three equally unsatisfactory. None of these three philosophical stances can soothe our feelings of alienation. Material monism alienates us from our deepest intuitions of nature and ourselves. Mental monism estranges us from our scientific aspirations and outlook. And material/mental dualism sets us apart from nature. Both monism and dualism arose in the context of the growth and success of mathematics and physics, but neither monism nor dualism was able to deal successfully with the countless philosophical oppositions that originated from the scientific production of abstractions—lifeless versus alive, recurrent versus creative, mechanical versus organic, external versus internal, meaningless versus meaningful, primary versus secondary and tertiary, essential versus non-essential, real versus illusory, objective versus subjective, *in itself* versus *for us*, natural versus supernatural, matter versus mind, fact versus value, theory versus practice, other-determined versus self-determined, and so on. Material monists try to explain away the second term in each of these oppositions, and mental monists, the first. By only looking at one side of each opposition, they all degenerate into one-eyed philosophers. Dualists, contrary to monists, are two-eyed philosophers, but unfortunately their "patchwork procedure" (NL 16) does not succeed in merging the two given images into one three-dimensional image.[2] In other words, they incorporate both sides of all oppositions in their account of reality, but fail to digest them and, hence, to deliver a coherent account. Both monists and dualists fail to turn all these alienating oppositions into coherent contrasts.

Three alternative modes of philosophizing, none satisfactory—the situation seems hopeless. There is, however, a fourth alternative, namely, the rejection of *both* the uncritical acceptance of all scientific abstractions as if they were the ultimate bricks to build a philosophical account of concrete reality, *and* the corresponding denial of all our deep intuitions as if they were necessarily inferior to our scientific abstractions. This fourth alternative rejects *both* the uncritical promotion of the concepts of mathematics and physics beyond their scientific action-radius as if they were "the ultimate categories of explanation" (FR 27), *and* the corresponding "pathological" demotion of the deep intuitions of mankind as if they were nothing but illusions (FR 11). The fourth alternative endeavors to outline a philosophy "capable of bridging the gap between science and that fundamental intuition of mankind which finds its expression in poetry, and its practical exemplification in the presuppositions of daily life" (SMW 95). The fourth alternative does not merely embrace scientific theories, while failing to include the scientist's intuitive practices. On the contrary, it holds: "Whatever is found in 'practice,' must lie within the scope of the metaphysical description. When the description fails to include the 'practice,' the metaphysics is inadequate and requires revision" (PR 13). The fourth alternative, contrary to the two monist and the dualist alternatives, first *assembles*[3] and then *transforms* the multiplicity of scientific abstractions and intuitive beliefs in opposition *into* a philosophical unity of coherent science-intuition contrasts (cf. PR 348). This fourth alternative is Whitehead's alternative, and his mode of philosophizing is the one that is also adopted by the nine authors of this anthology.

"The enormous success of the scientific abstractions," Whitehead wrote, "has foisted onto philosophy the task of accepting them as the most concrete rendering of fact" and, he added:

> Thereby, modern philosophy has been ruined. It has oscillated in a complex manner between three extremes. There are the dualists, who accept matter and mind as on an equal basis, and the two varieties of monists, those who put mind inside matter, and those who put matter inside mind. But this juggling with abstractions can never overcome the inherent confusion introduced by the ascription of misplaced concreteness to the scientific scheme. (SMW 55)

Whitehead's alternative is fighting "the Fallacy of Misplaced Concreteness"—the "error of mistaking the abstract for the concrete"—because "this fallacy is the occasion of great confusion in philosophy" (SMW 51). The fallacy of misplaced concreteness is committed each time abstractions are taken as concrete facts, and "more concrete facts" are expressed "under the guise of very abstract logical constructions" (SMW 50–51). This fallacy lies at the root of the modern philosophical confusions of scientific materialism, progressive bifurcation of nature, materialist monism, idealist monism, and the whole spectrum of magical, ontological (Cartesian), and epistemological (Kantian) dualisms. Indeed, the scientific method resulted in successful abstractions that were subsequently mistaken by philosophers as concrete facts, whereas the concrete facts revealed by our intuition were explained—and most often, explained away—in terms of the scientific abstractions to which concreteness was ascribed. But, wrote Whitehead: "The explanatory purpose of philosophy is often misunderstood. Its business is to explain the emergence of the more abstract things from the more concrete things. . . . In other words, philosophy is explanatory of abstraction, and not of concreteness" (PR 20). Moreover: "Philosophy destroys its usefulness when it indulges in brilliant feats of explaining away" (PR 17). Clearly, fighting the fallacy of misplaced concreteness does not mean fighting science, but getting philosophy back on track.

Whitehead challenged scientific materialism "as being entirely unsuited to the scientific situation at which we have now arrived," and yet, according to him, "scientific materialism . . . is not wrong, if properly construed," and he continued:

> If we confine ourselves to certain types of facts, abstracted from the complete circumstances in which they occur, the materialist assumption expresses these facts to perfection. But when we pass beyond the abstraction, either by more subtle employment of our senses, or by the request for meanings and for coherence of thoughts, the scheme breaks down at once. The narrow efficiency of the scheme was the very cause of its supreme methodological success. For it directed attention to just those groups of facts which, in the state of knowledge then existing, required investigation. (SMW 17)

> The advantage of confining attention to a definite group of abstractions, is that you confine your thoughts to clear-cut definite things, with clear-cut definite relations. Accordingly, if you have a logical head, you can deduce a variety of conclusions respecting the relationships between these abstract entities. Furthermore, if the abstractions are well-founded, that is to say, if they do not abstract from everything that is important in experience, the scientific thought which confines itself to these abstractions will arrive at a variety of important truths relating to our experience of nature. (SMW 58)

> The disadvantage of exclusive attention to a group of abstractions, however well-founded, is that, by the nature of the case, you have abstracted from the remainder of things. In so far as the excluded things are important in your experience, your modes of thought are not fitted to deal with them. You cannot think without abstractions; accordingly, it is of the utmost

importance to be vigilant in critically revising your modes of abstraction. It is here that philosophy finds its niche as essential to the healthy progress of society. (SMW 59)

I hold that philosophy is the critic of abstractions. Its function is the double one, first of harmonising them by assigning to them their right relative status as abstractions, and secondly of completing them by direct comparison with more concrete intuitions of the universe, and thereby promoting the formation of more complete schemes of thought. It is in respect to this comparison that the testimony of great poets is of such importance. Their survival is evidence that they express deep intuitions of mankind penetrating into what is universal in concrete fact. Philosophy is not one among the sciences with its own little scheme of abstractions which it works away at perfecting and improving. It is the survey of the sciences, with the special object of their harmony, and of their completion. It brings to this task, not only the evidence of the separate sciences, but also its own appeal to concrete experience. (SMW 87)

The creation of a scientific or philosophical masterpiece is often triggered by a sense of dissatisfaction with incoherencies in inherited doctrines. The masterpiece can then be characterized as the creative response of a genius to an assault on his longing for coherence. The creation of the special theory of relativity, for example, was triggered by Einstein's gnawing dissatisfaction with an incoherence in Maxwell's theory of electricity and magnetism, manifest in all cases where this theory leads not to a unique description, but to several reference-frame dependent descriptions for one and the same situation. Likewise, the early development of Whitehead's thought can best be understood in terms of a gradually growing feeling of dissatisfaction with the incoherence in philosophy that resulted from Newton's scientific materialism, Descartes' matter/mind dualism, Locke's primary/secondary bifurcation of nature, Hume's and Kant's theories of causality, Hegel's idealist monism, etc. Of course, feelings of dissatisfaction do not include the inspiration to overcome them. Einstein could not have created special relativity without his study of Mach and Poincaré—I limit myself to just two of Einstein's major sources of inspiration. Whitehead could not have created process philosophy without his study of Maxwell in physics[4] and Bergson and James in philosophy—again, this is a coarse-grained selection.[5]

It is Maxwell's theory of electricity and magnetism—even more than the theories of relativity and quantum mechanics—to which Whitehead referred when writing in 1925 about "the scientific situation at which we have now arrived," the situation, that is, with respect to which scientific materialism is "entirely unsuited" (SMW 17). This means that, according to Whitehead, scientific materialism is a very *inadequate* scientific cosmology. It *fails* to integrate not only the abstractions of the sciences of, for example, biology, physiology, and psychology, plus our deep intuitions of concrete reality, but even the theory of electromagnetism as initiated with Maxwell's 1873 *Treatise*—a theory that is part of the alleged "materialist" science of mathematical physics. The fact of being inspired by new theories of mathematical physics itself, in order to bridge the modern gap in philosophy between science and intuition (of course, while preventing the fallacy of mistaking the new abstractions for concrete facts), is typical not only for Whitehead, but also for several of the authors of this anthology.

In what follows, like Whitehead in *Nature and Life*, I limit the list of the inherited doctrines provoking Whitehead's dissatisfaction to Newton's "scientific materialism" (SMW 17) and Hume's "sensationalist empiricism" (PR 57). Also, I limit the developments that inspired Whitehead's creation of process philosophy—also called "organic" philosophy—to the development of Maxwell's theory and to the emergence of a "radical empiricism" in the writings of a spectrum of philosopher-psychologists ranging from Bergson in France to James in America. I am justified in my limitation to psychology and electromagnetism as Whitehead's main sources of inspiration by two of his own claims. One: "If you start from the immediate facts of our psychological experience, as surely an empiricist should begin, you are at once led to the organic conception of nature" (SMW 73). Two: "It is equally possible to arrive at this organic conception of the world if we start from the fundamental notions of modern physics . . . By reason of my own studies in mathematics and mathematical physics, I did in fact arrive at my convictions in this way. Mathematical physics presumes in the first place an electromagnetic field of activity pervading space and time" (SMW 152).

## NATURE AND LIFE

"Modern physical science," Whitehead wrote in his 1934 booklet, *Nature and Life*, "is the issue of a coordinated effort, sustained for more than three centuries, to understand those activities of Nature by reason of which the transitions of sense-perception occur" (65). But according to Whitehead, Hume's sensationalist empiricism has undermined the idea that our perception can reveal those activities, and Newton's scientific materialism has failed to render his formulae of motion and gravitation intelligible.

Whitehead was dissatisfied with Hume's reduction of perception to sense perception because, as Hume discovered, pure sense perception reveals a succession of spatial patterns of impressions of color, sound, smell, etc. (a procession of forms of sense data), but it does not reveal any causal relatedness to interpret it (any form of process to render it intelligible). In fact, all "relatedness of nature," not only its causal relatedness, was "demolished by Hume's youthful skepticism" (R 13) and conceived as the outcome of mere psychological association. Whitehead wrote: "Sense-perception, for all its practical importance, is very superficial in its disclosure of the nature of things. . . . My quarrel with [Hume] concerns [his] exclusive stress upon sense-perception for the provision of data respecting Nature. Sense-perception does not provide the data in terms of which we interpret it" (NL 21).

Whitehead was dissatisfied with Newton's conception of nature as the succession of instants of spatial distribution of bits of matter for two reasons. One: the concept of a "durationless" instant, "without reference to any other instant," renders unintelligible the concepts of "velocity at an instant" and "momentum at an instant" as well as the equations of motion involving these concepts (NL 47). Two: the concept of self-sufficient and isolated bits of matter, having "the property of simple location in space and time" (SMW 49), cannot "give the slightest warrant for the law of gravitation" that Newton postulated (NL 34). Whitehead wrote: "Newton's methodology for physics was an overwhelming success. But the forces which he introduced left Nature still without meaning or value. In the essence of a material body—in its mass, motion, and shape—there was no reason for the law of gravitation" (NL 23). "There is merely a formula for succession. But there is an absence of understandable causation for that formula for that succession" (NL 53–54).

"Combining Newton and Hume," Whitehead summarized, "we obtain a barren concept, namely, a field of perception devoid of any data for its own interpretation, and a system of interpretation devoid of any reason for the concurrence of its factors" (NL 25). "Two conclusions," Whitehead wrote, "are now abundantly clear. One is that sense-perception omits any discrimination of the fundamental activities within Nature. . . . The second conclusion is the failure of science to endow its formulae for activity with any meaning" (NL 65). The views of Newton and Hume, Whitehead continued, are "gravely defective. They are right as far as they go. But they omit . . . our intuitive modes of understanding" (NL 26).

In Whitehead's eyes, however, the development of Maxwell's theory of electromagnetism constituted an antidote to Newton's scientific materialism, for it led him to conceive the whole universe as "a field of force—or, in other words, a field of incessant activity" (NL 27). The theory of electromagnetism served Whitehead to overcome Newton's "fallacy of simple location" (SMW 49), that is, the conception of nature as a universe of self-sufficient isolated bits of matter. Indeed, we cannot say of an electromagnetic event that it is "here in space, and here in time, or here in space-time, in a perfectly definite sense which does not require for its explanation any reference to other regions of space-time" (SMW 49). The theory of electromagnetism "involves the entire abandonment of the notion that simple location is the primary way in which things are involved in space-time" because it reveals that, "in a certain sense, everything is everywhere at all times" (SMW 91). "Long ago," Whitehead wrote, Faraday already remarked "that in a sense an electric charge is everywhere," and: "The modification of the electromagnetic field at every point of space at each instant owing to the past history of each electron is another way of stating the same fact" (CN 148).

The lesson that Whitehead learned from the theory of electromagnetism is unambiguous:

> The fundamental concepts are activity and process. . . . The notion of self-sufficient isolation is not exemplified in modern physics. There are no essentially self-contained activities within limited regions. . . . Nature is a theatre for the interrelations of activities. All things change, the activities and their interrelations. . . . In the place of the procession of [spatial] forms (of externally related bits of matter, modern physics) has substituted the notion of the forms of process. It has thus swept away space and matter, and has substituted the study of the internal relations within a complex state of activity. (NL 35–36)

But overcoming Newton was insufficient for Whitehead because Hume "has even robbed us of reason for believing that the past gives any ground for expectation of the future" (NL 65). According to Whitehead, "science conceived as resting on mere sense-perception, with no other sources of observation, is bankrupt, so far as concerns its claims to self-sufficiency" (NL 66). In fact, science conceived as restricting itself to the sensationalist methodology can find neither efficient nor final causality. It "can find no creativity in Nature; it finds mere rules of succession" (NL 66). "The reason for this blindness," according to Whitehead, "lies in the fact that such science only deals with half of the evidence provided by human experience" (NL 66).

Contrary to Hume, Whitehead held that it is untrue to state that our perception, in which sense perception is only one factor, discloses no creative play of efficient and final causes (cf. NL 68). Inspired by the anti-associationism and radical empiricism of Bergson and James, Whitehead launched a new answer to the question for the proper analysis of perception. "The conventional answer to this question," he wrote, "is that we perceive Nature through our senses. Also, in the analysis of sense-perception we are apt to concentrate upon its most clear-cut instance, namely sight" (NL 74). However:

> In the first place, even in visual experience, we are also aware of the intervention of the body. We know directly that we see *with our eyes*. That is a vague feeling, but extremely important.

Second, every type of crucial experiment proves that what we see, and where we see it, depend entirely upon the physiological functioning of our body. (NL 75)

"Now," according to Whitehead, "the same (direct awareness of the causal efficacy of the body) is true of all other modes of sensation" (NL 76). Moreover, next to our intuition of derivation from our body, "our immediate experience also claims derivation from another source," namely, "our own state of mind directly preceding the immediate present of our conscious experience" (NL 78). "Thus," Whitehead concluded,

> our experience in the present discloses its own nature in two sources of derivation, namely, the body and the antecedent experiential functionings. Also, there is a claim for identification with each of these sources. The body is mine, and the antecedent experience is mine. Still more, there is only one ego, to claim the body and to claim the stream of experience [that constitutes the soul]. I submit that we have here the fundamental basic persuasion on which we found the whole practice of our existence. (NL 79-80)

An analysis of Whitehead's concept of intuition is given in Chapters Three and Four of this anthology, but it is important to already notice that when Whitehead speaks about "direct awareness," "immediate experience," "basic persuasion," etc., we might substitute "intuition." If we do so in the next quote, it offers a synopsis of where we are:

> It is the task of philosophical speculation to conceive the happenings of the universe so as to render understandable the outlook of physical science and to combine this outlook with these direct persuasions representing the basic facts upon which [a radical empiricist] epistemology must build. The weakness of the [sensationalist] epistemology . . . was that it based itself purely upon a narrow formulation of sense-perception. Also, among the various modes of sensation, visual experience was picked out as the typical example. The result was to exclude all the really fundamental factors constituting our experience. (NL 83–84)

In order to include these factors, Whitehead asked his readers to consider the fundamental interconnections of "body and soul, of body and Nature, of soul and Nature," and to notice that they share a "very remarkable characteristic" (NL 84), namely:

> There is a dual aspect to the relationship of an occasion of [the stream of] experience [constituting the soul] as one relatum and the experienced world as another relatum. The world is included within the occasion in one sense, and the occasion is included in the world in another sense. . . . [And] this baffling antithetical relation extends to all the connections which we have been discussing. (NL 85–86)

> We are in the world and the world is in us. Our immediate occasion is in the society of occasions forming the soul, and the soul is in our present occasion. The body is ours, and we are an activity within our body. This fact of observation, vague but imperative, is the foundation of the connexity of the world. (NL 89–90)

This imperative intuition discloses that "the togetherness of things involves some doctrine of mutual immanence. In some sense or other, this community of the actualities of the world means that each happening is a factor in the nature of every other happening" (NL 87). Hume demolished the relatedness of nature; Whitehead restored it. For example, he founded the "doctrine of causation . . . on the doctrine of immanence," and wrote: "Each occasion presupposes the antecedent world as active in its own nature. . . . This is the doctrine of causation" (NL 88–89).

After a more extensive account than I can offer here, Whitehead brought *Nature and Life* to its final conclusion: "In this survey of the observational data in terms of which our philosophic cosmology must be founded, we have brought together the conclusions of physical science, and those habitual persuasions dominating . . . mankind" (NL 90). In other words, in his post-Humean assemblage of the data to found his post-Newtonian process cosmology, Whitehead has brought together the abstract concepts of Maxwell's mathematical physics and the deep intuitions of humankind.

## INTUITION IN MATHEMATICS AND PHYSICS

According to the Dalai Lama: "We need a worldview grounded in science that does not deny the richness of human nature and the validity of modes of knowing other than the scientific."[6] The process cosmology that Whitehead presented in his 1929 book, *Process and Reality*, is just such a worldview. The 1920s, however, was the era of Russell's logical atomism and the birth of logical positivism. Russell and the logical positivists embraced Hume's philosophical heritage, and they rejected Whitehead's alternative mode of philosophizing.

At first, Russell and the logical positivists placed great value on Whitehead's philosophical writings because Whitehead's early writings pivoted around the method of extensive abstraction. Whitehead developed this method to avoid the fallacy of misplaced concreteness in the philosophy of geometry. Indeed, instead of trying to explain our concrete intuition of spatial (resp. temporal) extension in terms of abstract points (resp. instants) to which misplaced concreteness was ascribed, his method intended to construct (by abstraction) the latter from the former in terms of the mathematical logic of classes and relations that Russell and Whitehead jointly developed and applied in *Principia Mathematica*. For Russell, the method of extensive abstraction was a paradigm of what his logical atomism was all about—to extend the scope of logical construction from the concepts of mathematics to those of physics, hence transforming his logical philosophy of mathematics (his logicism) into a logical philosophy of science as a whole (mathematics, physics, etc.). Russell, however, endeavored to construct the concepts of physics not from vague intuitions, but from clear atoms, namely, sense data and logical truths, hence completely ignoring the anti-Hume, pro-Bergson, pro-James stance of Whitehead.[7] For Whitehead, the method was also the way to develop and present—in 1919 and 1920 (in PNK and CN)—a philosophical alternative to Einstein's interpretation of special relativity. Einstein (mistakenly) ascribed concreteness to abstractions (such as the constancy of the speed of light) in order to derive the structure of space-time. Conversely, Whitehead started from our concrete intuitions of spatio-temporal extension, of motion and rest,[8] and then abstracted the structure of space-time.

However, when Whitehead made the additional step in 1922 (R) of offering a philosophical alternative to Einstein's interpretation of general relativity, Russell and the logical positivists unambiguously favored Einstein's interpretation and rejected Whitehead's. Instead of ascribing (misplaced) concreteness to the abstract space-time structure of physics, and instead of describing the more concrete physical phenomena of gravitation in terms of how this alleged concrete space-time structure curves in the presence of matter, which is what Einstein did, Whitehead argued differently. He explained these phenomena by conceiving the field of gravitation as completely analogous to the field of electromagnetism—the abstract concept he felt harmonized with our deep intuitions of internally related activities, causality, etc., as highlighted in the second part of this introduction.[9] Whitehead's belief in internal relations and causality was an insult to the anti-metaphysical dogmas of Russell; his reliance on our intuition of the metric relatedness of nature was at variance with Poincaré's conventionalism; and his method of causal explanation was an antipode of Mach's method of economic description. In short, Whitehead distanced himself from the idols of the logical positivists: Hume, Mach, Poincaré, Einstein, and Russell.[10] Contrary to Russell, Whitehead shied away from public controversy, so that in his early philosophical writings (OT, PNK, CN, and R) his anti-positivist stance was not all that clear to his contemporaries.

However, when Whitehead finally made public his alternative mode of philosophizing in 1925 (SMW), and his alternative scientific cosmology in 1929 (PR), Russell and the logical positivists could no longer see Whitehead as a philosopher of the future, and they could no longer think of his philosophy as a paradigm of logical and analytical thinking. Consequently, Whitehead was rejected as a philosopher of the past, and his philosophy was stigmatized as an anachronistic and obscure metaphysical system.

Today, logical positivism belongs to the past, but this does not imply that Whitehead's relevance for the present has been broadly recognized. And despite all revolutions in science and philosophy, Newtonian scientific materialism and Humean skepticism are still lingering in the background of our contemporary mentality. Consequently, the idea of a necessary gap between science and intuition is reconfirmed again and again; and the corresponding idea of a war between science and intuition, in which each achievement of science is marketed as a defeat of intuition, is more dominant than ever.

To give but one example, in his 2012 book, *Mind and Cosmos*, Thomas Nagel relied on the compass of human intuition to conclude that "the materialist neo-Darwinian conception of nature is almost certainly false." This book has been greeted by a storm of extremely negative reviews, and in H. Allen Orr's review, "Awaiting a New Darwin," we can read one of the major reasons why the reception of Nagel's book was so negative.[11] Orr writes:

> There's not much of an argument here. Instead Nagel's conclusion rests largely on the strength of his intuition. His intuition recoils from the claimed plausibility of neo-Darwinism and that, it seems, is that. . . . But plenty of scientific truths are counterintuitive . . . and a scientific education is, to a considerable extent, an exercise in taming the authority of one's intuition.

Instead of endorsing the modern dogma that a truth cannot be scientific unless it hurts the deep intuitions of mankind, and that we cannot be scientific unless we tame the authority of our intuition, the spirit of this anthology is to resist philosophical claims that do violence to what we cannot help but believe, to question the authority of science-inspired philosophers to disenchant and reduce nature and humankind in the name of successful but limited scientific abstractions, and to stress the contemporary relevance of Whitehead's mode of philosophizing and issuing process cosmology.

In *Chapter One*, Ronald Phipps offers a compelling alternative to disenchanting and reductionist modes of philosophizing, namely, an 'integral' or 'integrative' mode of philosophizing inspired by Whitehead.

Whitehead always avoided public controversy, but in a private letter written to Henry S. Leonard in 1937—a letter that has been made public thanks to Leonard's son (Henry S. Leonard, Jr.) and the efforts of Phipps himself—Whitehead unambiguously rejected logical positivism. Phipps' essay first offers the reader the most revealing quotes from this letter, then adds a selection of Leonard's comments on it. This sets the stage for the remainder of his essay, in which Phipps first elaborates on what integral philosophy is, in contrast with logical positivism and analytic philosophy, then highlights the critical potential and added value of integral philosophy in the domains of space-time geometry, cosmology, and particle physics.

In his account of space-time geometry, Phipps stresses the importance not only of Whitehead's method of extensive abstraction, but also of the related calculus of individuals, developed by Leonard and Nelson Goodman, and of the analogous theory of sponges developed by Karl Menger. In his account of cosmology, Phipps launches a Whitehead-inspired attack on the myth of the big bang. And finally, in his account of particle physics, Phipps aims at showing the fruitfulness of Whitehead's event-ontology in comparison with both the standard model and string theory.

*Chapter Two*, in a sense, starts same way. Phipps' essay starts with Leonard — Phipps was once his personal assistant — with the aim of contrasting logical positivism and integral philosophy. Timothy Eastman's essay starts with Herbert Feigl, one of the founders of logical positivism — Eastman was once his student — with the aim of contrasting Feigl's reductionist, nothing-but-philosophy, with Whitehead's speculative, something-more-philosophy. For Eastman, Feigl exemplified the contradiction of scientifically and explicitly denying the existence of (i) feelings, (ii) final causes, and (iii) values, while intuitively and implicitly affirming them; the contradiction, that is, of limiting reality in theory to (i) substances, (ii) efficient causes, and (iii) sheer facts, while continually transgressing these limitations in the practice of day-to-day life. As Whitehead once wrote: "Scientists animated by the purpose of proving that they are purposeless constitute an interesting subject for study" (FR 16).

For Whitehead, the substance-predicate structure of language is the result of an elaborate history of abstraction, and the idea that this structure is a direct reflection of the deep structure of reality, an instance of the fallacy of misplaced concreteness. In the first part of his essay, inspired by the fact that substance-talk has become increasingly inadequate in microphysics, Eastman makes a plea to replace the traditional substance-ontology with Whitehead's process-ontology. In the second part, Eastman highlights several revolutionary notions — for example, the notions of "prehension" in the philosophy of Whitehead, "emergence" in the complex system dynamics of Stuart Kauffman, "logical causation" and "non-Boolean possibility space" in the quantum philosophy of Michael Epperson and Elias Zafiris, and "decision" in the freewill theorem of John Conway and Simon Kochen. Inspired by these notions, Eastman argues that the time is ripe for philosophers and scientists alike to overcome their fixation on sheer efficient causality. In the third part, Eastman lists a number of novel philosophical approaches — for example, the speculative naturalism of Arran Gare and the "inclusive" speculative approach of George R. Lucas, Jr., Nicholas Rescher, and George Shields, which includes the best of analytic philosophy, process philosophy, and science. The approaches that Eastman lists and endorses all aim at overcoming "scientism," that is, the common presumption that all meaningful propositions can be reduced to value-free, sheer factual scientific propositions.

Eastman's essay ends with his speculative suggestion that our intuition might be associated with the amplification of subtle quantum effects within our body, which tap both the "Boolean space" of what is actualized (i.e., of the past), and the "non-Boolean possibility space" of the near future.

The starting point of *Chapter Three* is Whitehead's concept of intuition, but in his essay Farzad Mahootian offers much more to the reader than a cut-and-dried account of this concept. Mathootian introduces intuition in the context of Whitehead's endeavor to "ecologically" re-envision science and philosophy, and to "rescue" this "type of thought," indebted to Bergson and James, "from the charge of anti-intellectualism" (PR xii). Mohootian uses the intricacies of Whitehead's categoreal scheme and his theory of prehensions to demonstrate that intuition, as conceived by Whitehead, is a kind of "rational intuition." In fact, intuitive judgments belonging to the third type that Whitehead distinguished — the "suspended" judgments constituting our "conscious imagination" — "are weapons essential to scientific progress" (PR 275).

In suspended judgments, propositions are entertained not because they are true or false, but because they are interesting. In fact, instead of joining narrow-minded logicians and moralists in their exclusive "preference for true propositions," and their habit of throwing false propositions "into the dust-heap" (PR 259), Mahootian embraces Whitehead's philosophical defense of false propositions by giving an especially compelling concrete example in the field of medical decisionmaking.

"The transitions to new fruitfulness of understanding," Whitehead wrote, "are achieved by recurrence to the utmost depths of intuition for the refreshment of the imagination" (AI 159), and Mathootian's essay ends with a fine example of such a transition, namely, the transition *from* the materialistic and mechanistic understanding of molecular biology, which contributes to the scientific alienation from nature, *to* the scientific and yet intuitive mimesis of the organic interrelatedness of nature in biomimetics, which has the potential to grow into a significant scientific contribution to an ecological civilization.

The starting point of *Chapter Four* is the same as the starting point of Chapter Three—Whitehead's concept of intuition. But whereas Mathootian's essay explores its implications for science and civilization in general, Ronny Desmet's short essay focuses on the role of intuitive recognition in mathematics and physics. Hence, Chapter Four represents a transition from the broad philosophical scope of the three chapters that precede it to the more specific topics of the six chapters that follow and that constitute the remainder of this anthology. Nonetheless, instead of exemplifying the notion of intuitive recognition with an intricate example from mathematics or physics, Desmet exemplifies it with an easy-to-understand *Gestalt* experiment.

Whitehead stressed that whatever is found in practice must lie within the scope of philosophy, and that whenever philosophy fails to include the practice, it is inadequate and requires revision. This might be the motto of the new current in the philosophy of mathematics, wherein the focus is primarily on mathematical practices instead of theories. Representatives of this new approach are, as are process philosophers, acutely aware of the primary importance of taking into account the concrete processes of creating and entertaining the abstract products of mathematics. One of the fathers of this new philosophy of mathematical practices is Jean Paul Van Bendegem, and one of the members of Van Bendegem's research center at the Free University of Brussels is Ronny Desmet. They are jointly responsible for Chapters Five and Six of this anthology. Both these chapters focus on beauty in mathematics, which addresses one of the most relevent elements in this new philosophy of mathematics; namely, to develop a philosophy that takes into account not merely the mathematical results as published in journals and books, but also the values that mathematicians experience in practice.

*Chapter Five* first offers a short account of Whitehead's philosophy of the aesthetic process of becoming, then checks its adequacy with respect to the concrete experience of mathematicians. This check reveals the appropriateness of Whitehead's philosophy to describe the mathematician's experience of beauty in mathematics. Desmet also shows that Whitehead's idea of the background-dependency of the emergence of beauty in an occasion of experience fits nicely with Gian-Carlo Rota's idea of the context-dependency of the experience of mathematical beauty. And finally, by putting the two-color sphere theorem in the context of the proof of the Kochen-Specker theorem in quantum mechanics, Desmet illustrates the truthfulness of Rota's aphorism: "The beauty of a theorem is best observed when the theorem is presented as the crown jewel within the context of a theory."

In *Chapter Six*, Van Bendegem and Desmet focus on the complementary aesthetic experiences that can accompany (i) the understanding of mathematical proofs and (ii) the creative process of innovative mathematical research. Following François Le Lionnais, they speak of (i) the "classical" beauty of mathematical proof, and (ii) the "romantic" beauty of mathematical creation. Their detailed account — which also invokes the views of three famous mathematicians: Birkhoff, Hardy, and Poincaré[12] — leads them to the following synoptic conclusion (see 143):

> Whitehead does not have a separate classical aesthetics of mathematical proof, but his aesthetics of experience can account for the aesthetic delight in proof, which emerges from the return of intuitive enlightenment on investment of

logical effort in proof. Also, Whitehead does not have a separate romantic aesthetics of mathematical creation, but his aesthetics of experience can account for the romantic and adventurous explorations of the mathematical landscape, which lead mathematicians from one mountain top to another, from one panoramic view to another, from one type of order to another, and from aesthetic delight to aesthetic delight, but not without descending in the valleys of discord, not without suffering the loneliness of non-conformism, not without venturing along the borders of chaos, and not without exercising the challenging freedom that can save us from our smallness and lack of vision.

In *Chapter Seven*, Arran Gare holds that in order to promote an ecological civilization we are in need of a new type of mathematics, a mathematics of becoming, which replaces the traditional type, the mathematics of being.

Traditional mathematics has flourished as the tool *par excellence* to explain away the "appearance" of nature as alive, creative, and organic; in other words, to explain away the "illusion" of becoming. It has focused exclusively on lifeless, recurrent, and mechanical forms, so that it came to be identified with a science of being rather than becoming. Especially via physics and philosophy, it has contributed to the bifurcation of nature, to our alienation from nature and from ourselves, and to the ecological catastrophe that humankind is facing today.

Gare's essay first puts the diagnosis of the alienating tendency of traditional mathematics in terms of Nietzsche's work. Nietzsche noted: "To impose upon becoming the character of being—that is the supreme will to power," and he held that the will to power, turned against itself, brought about the nihilism of European civilization (its denial of meaning and value in life). As nothing contributed more to imposing upon becoming the character of being than traditional mathematics and the physics it inspired, and as "the method of philosophy has also been vitiated by the example of mathematics" (PR 10), the nihilistic tendency of traditional mathematics is at once evident. And according to Gare, it is nihilism that lies at the root of the ecological catastrophe that we are facing today.

Whitehead signaled the emergence of a new mathematics when he wrote that "the new physics" (read: Maxwell's electromagnetism and the theories of relativity as conceived by Whitehead) has substituted the notion of "forms of process" in place of the notion of the "procession of forms" (NL 36). Instead of trying to understand nature as a procession of mathematical patterns of being, Whitehead promoted the understanding of nature in terms of mathematical patterns of processes of becoming.

Gare first shows that Whitehead's mature philosophy of mathematics has more in common with the intuitionism of Poincaré, or Brouwer, or Weyl, than with the logicism of Frege and Russell, or the formalism of Hilbert. However, Gare adds that contrary to the former three, but similar to Peirce, Whitehead was led away from formalism and logicism towards intuitionism not by the constructivism of Kant, but by the more radical constructivism of Schelling, which was operative in the background of one of Whitehead's major sources of inspiration: Hermann Grassmann's *Ausdehnungslehre* (UA x).

Gare then shows that Whitehead's intuitionist conception of mathematics is a precursor of the conception that goes with category theory. The latter has been further developed by relational systems biologist Robert Rosen into a general theory of modeling, which has been improved by Louie, Ehresmann, Vanbremeersch, etc., and which has stimulated further efforts to develop a new mathematics that is adequate to model nature alive, creative, and organic; for example, the non-reductionist biomathematics, or "integral biomathics" of Simeonov.

Gare's essay ends by stressing that with this new conception of mathematics we can now view mathematics as playing a major part in comprehending a creative universe of processes of becoming, rather than explaining away creativity and the reality of life. Fully developed, according to Gare, such mathematics could be an important factor in advancing an ecological civilization that augments life rather than undermining it.

*Chapter Eight* has three parts. In the first part of his essay, Gary Herstein admits feeling uncomfortable with the word "intuition." One of the elements provoking Herstein's discomfort is the appeal to a mysterious faculty of intuition in response to Eugene Wigners' claim of a mysterious fit between our mathematical theories and external reality. Herstein can only accept using the term "intuition" if it is conceived of in a radically empiricist way or, better, by appealing to Whitehead's version of James's radical empiricism. So Herstein does not agree to explain Wigner's claim by postulating a special faculty of intuition at the base of our study of mathematical patterns which mysteriously pre-establishes harmony between these internal, ideal patterns and the external, real patterns of relatedness in nature. According to Herstein, there is no mystery to start with, for there is no initial gap between some "internal experience" responsible for our mathematical creations, and some "external reality" that we successfully model with them. Indeed, by conforming to Whitehead's understanding of radical empiricism, we can claim that our experience is part of the real, and that relations, like the related entities themselves, are directly experienced (prehended or felt) so that the totality of reality is part of our experience. Our mathematical intuition is not a separate faculty, but a subset of the set of intuitive judgments ("insights," Herstein writes) which emerge from the largely unconscious concrescence of prehensions or feelings of the real, and which constitute what we call "human intuition." The fact that intuitive judgments are intellectual feelings, hypothetical and subject to further observation and reasoning, saves Whitehead's notion from the anti-intellectualism that Herstein mistrusts in appeals to some mysterious faculty of intuition.

In part two of his essay, Herstein talks about Whitehead's theory of gravity. Contrary to Einstein's theory of gravity—that is, the general theory of relativity—Whitehead's theory of gravity observed coherence with our intuition of the uniformity of nature, i.e., our direct experience of the uniform relatedness of nature that forms the subject-matter of space-time geometry, and which Whitehead distinguished from the contingent relatedness of nature that forms the subject-matter of field physics. Contrary to Einstein, Whitehead also respected the "logical" necessity ("logic" conceived here as "the theory of inquiry") of establishing geometrical relations prior to establishing "meaningful" physical measurements.

In Einstein's theory, geometrical and physical relations are collapsed, and the space-time (or geo-metrical) tensor expressing the former relations is identified with the field (or physico-metrical) tensor expressing the latter. Hence it is a mono-metric theory, incoherent with our intuition, and leaving "the whole antecedent theory of measurement in confusion" (R 83). Contrary to Einstein's theory, Whitehead's is a bimetric theory. In fact, it is the first bimetric theory of gravity (published 18 years prior to Nathan Rosen's alleged first bimetric theory), and it justifies not one particular bimetric equation of gravity, but a whole family of such equations, some of which might be even closer to Einstein's field equation qua empirical predictions than the one Whitehead preferred (R 83).

Herstein thinks that the success of Einstein's monometric theory, despite Whitehead's critique and bimetric alternative, involves both the aesthetic appeal of Einstein's intuitions and the fact that a fit between Einstein's theory and observation was manufactured. Indeed, Einstein also invoked intuition but, as Herstein makes clear, Einstein did not rely on direct and concrete intuitions experienced in day-to-day practice, but promoted educated and abstract intuitions, dominated by theory. And indeed, the theory-observation fit was manufactured, in the sense that freely adjustable parameters have been added to Einstein's theory, which guarantee upfront such a fit. (Herstein holds that we should not exaggerate, as Wigner does, the success of mathematics to model reality, for some of it is manufactured.) The third part of Herstein's essay can be read as a warning against giving abstract intuitions priority over concrete ones and against giving theoretical models priority over empirical facts—an unfortunate strategy that is followed, for example, by most cosmologists and string theorists.

Presentations of quantum mechanics often stress the counter-intuitive nature of its most salient results; for example, the violation of the Bell inequalities and the verification of the Kochen-Specker theorem. It is possible, however, to bring science and intuition closer together here, as well. It is, for example, worth mentioning the 2010 paper of Harald Atmanspacher and Thomas Filk,[13] in which, inspired by the mathematics of a two-state quantum Zeno effect, a mathematical model is proposed that describes the effective dynamics of switching mental states during our perception of ambiguous figures like the Necker cube and the duck-rabbit figure, and giving rise to a violation of *temporal* Bell inequalities. Atmanspacher and Filk transpose an apparently counter-intuitive result in the domain of quantum mechanics to the domain of the direct experience of bistable perception. It is also worth mentioning the intuitive analogy, presented in Chapter Five, between the emergence of the property of spin in a quantum spin-measurement according to the Kochen-Specker theorem and the emergence of the quality of beauty in human experience. A third example is the starting point of *Chapter Nine*.

From the nonlocal entanglement of elementary particles (giving rise to the violation of the Bell inequalities) and the particularities of spin-measurement (giving rise to the Kochen-Specker theorem), John Conway and Simon Kochen derived their freewill theorem. In his essay, Robert Valenza first refers to the derivation of the Conway-Kochen freewill theorem as given in his paper "Possibility, Actuality, and Freewill"—yet another example of a simple and intuitive approach to a complex and abstract original. Moreover, with the freewill theorem itself, Conway and Kochen also seem to bring intuition and mathematical physics closer together. Indeed, they interpret their theorem as implying that (in Valenza's words) "if humans have freewill, then so do, say, photons." However, their notion of freewill (Valenza immediately adds) "is a very peculiar notion of freewill that has nothing to do with the ordinary modality of intention." Instead of speaking of freewill, it might be better to say that both humans and, say, photons are "undetermined *by all the information available to them in the universe.*"

With this last phrase, Valenza introduces the main topic of his essay: the concept of "information," which has become crucial in microphysics as well as in many other branches of science, such as communications theory. Valenza identifies three essential aspects of information, corresponding to the notions of *physical substrate*, *abstract structure*, and *semantic efficacy*, each of which is intuitively and yet precisely treated in a separate section of his essay.

Valenza holds that information involves the *perception* of definite contrasts hosted in a *physical substrate.* Hence lack of contrasts (mere uniformity) implies lack of information; binary contrasts imply the link of Valenza's concept of information with information theory; and, the contrasts hosted in a physical substrate can be conceived of as the potential information processed into information by human perception or some nonhuman process. (The potential information that is part of a CD is processed into information by the CD-player playing it, even if no one is listening.)

Valenza holds that the processing of potential information into information also involves the abstraction of form from the physical substrate, which is a particular instance of the abstracted form. The detachment of the *abstract structure* is part of the process in which information emerges, and reminds Valenza of the role of physical recognition and conceptual reversion in Whitehead's analysis of process. Valenza also links the "perspective invariance" of the abstract structure, which makes it "objective," with the notions of "covariance" in Einstein's general theory of relativity and in category theory.

Next, the processing of potential information into information involves the reference of the abstract structure to the *semantic content*, but it is important to notice that this does not imply a simple (symbol-meaning) *reference* from the abstract structure back to the bearer of potential information, the physical substrate. The *semantic efficacy*, which Valenza also terms the *in-formation*, goes beyond the efficacy of a simple reference. For example, the processing of the potential information of a particular one-dollar bill into information involves the semantic efficacy not only of the symbolic reference of the one-dollar bill pattern to the particular bill, but also of the total social context, which we call the monetary system, and which "in-forms" the particular one-dollar bill that is being processed.

Valenza's paper ends with the application of his concept of information to some issues in the philosophy of physics. I limit myself to one remark: Valenza holds that, when we try to get a grip on the quantum entanglement of particles in some measurement-contexts, we do well to also rely on the notion of semantic efficacy and its limitations instead of merely relying on the notion of causal efficacy and its contingent limitation to the speed of light in vacuum. According to Valenza, it is intuitively more satisfying to also speak of the context-dependent possibility or impossibility to "in-form," than to merely speak of the alleged context-independent impossibility to causally transmit information at a speed faster than that of light in vacuum.

*Chapter Ten* is the longest and most technical of the ten chapters in this anthology, but it is easier to summarize than most. Jesse Bettinger's essay explores the possibility of giving a neurobiological account of intuition. By identifying "intuitions" with "gut feelings," Bettinger, in the first part of his essay, can draw on the neurobiological account of the faculty of "interoception"—the ability to sense the internal conditions of the body—which focuses on the neurobiological architecture and dynamics of the brain-gut information pathway (the "neurovisceral axis" in which "von Economo neurons" facilitate the information flow). In the second part of his essay, Bettinger then correlates key features of the neurobiological account of intuition with major concepts of Whitehead's process philosophy. The neurobiological dynamic is correlated to the process dynamic as outlined in Whitehead's theory of prehensions, and as particularized in his theory of human perception (involving three modes of perception: causal efficacy, presentational immediacy, and symbolic reference). Inspired by the connection he established between the neurobiological account and Whitehead's process philosophy, as well as by Ralph Pred's 2009 book, *Onflow*, Bettinger is prompted to review the bottom-up information processing of the neurobiological account from a mere additive accumulation to a selective synthesis of values. This shows that the connection is a symbiotic one, leading to cross-fertilizations that can advance both the modern neuroscience of interoception and the philosophy of process.

To conclude, no matter how diverse the ten essays of this anthology may seem at first glance, this introduction hopefully makes clear that they have much in common. All authors share an urge for coherence. That is why they all share the concern to overcome the modern habit of interpreting and presenting new results in mathematics and physics as promoting a scientific worldview that does not cohere with, but is opposed to, our deepest intuitions of ourselves and of nature. Moreover, all authors share an awareness of the links between the modern science/intuition bifurcation of nature and the monist as well as dualist strategies of philosophy; between these strategies and our alienation from ourselves and from nature; and between our alienation and the ecological catastrophe we are facing today, involving the decrease of natural resources, pollution, climate change, etc. Consequently, all authors share a sense of urgency and hold that the time has come to counter the strategy of interpreting and presenting each success of science as one more step in the demonstration of the overall failure of our intuition to guide science, i.e., one more step in the disenchantment of ourselves and the world of which we are part. The essays of this anthology have been written with the common aim of countering this reductionist and nihilistic strategy by advancing Whitehead's philosophical research program of *thinking things together* — science and intuition, facts and values, etc. — to promote the fundamental coherence that is required to start building an ecological civilization.

## NOTES

1    Whitehead, however, did not follow Berkeley in holding that "the boat" wherein all qualities are held is the mind of God. For Whitehead it was the process of becoming. Whitehead rejected both scientific materialism and Berkeley's idealism. In fact, Whitehead rejected all "philosophical idealism which finds the ultimate meaning in mentality that is fully cognitive," because "this idealistic school . . . has been too much divorced from the scientific outlook" (SMW 63). "For Berkeley's mind," Whitehead wrote, "I substitute a process of prehensive unification" (SMW 69). This is not the place to introduce Whitehead's process philosophy, but here is just one more relevant quote: "The reality is the process. It is nonsense to ask if the colour red is real. The colour red is ingredient in the process of realization" (SMW 72). Note the analogy with the primary quality of "spin" and the tertiary quality of "beauty" as presented in Chapter Five of this anthology.

2    According to Whitehead, both monism and dualism are characteristic of a "one-eyed reason, deficient in its vision of depth" (SMW 59).

3    "Philosophy can exclude nothing. Thus it should never start from systematization. Its primary stage can be termed *assemblage*" (MT 2).

4    For details on Whitehead's study of Maxwell's theory of electricity and magnetism, see Ronny Desmet, "Whitehead's Cambridge Training," in *Whitehead: The Algebra of Metaphysics*, edited by Ronny Desmet and Michel Weber, 91–125 (Louvain-la-Neuve, Belgium: Les éditions Chromatika, 2010).

5    For a more refined selection of Whitehead's sources of inspiration, see Ronny Desmet, "The Gestalt Whitehead" *Process Studies* 44.2 (2015): 190–223.

6    Online editorial review, *War of the Worldviews: Where Science and Spirituatily Meet—and Do Not,* by Deepak Chopra and Leonard Mlodinow.

7    Russell also ignored the critical function Whitehead ascribed to philosophy. Russell's logical atomism, like logical positivism, transformed philosophy into a handmaiden of science. For details on the divergence of Russell from Whitehead, see Ronny Desmet, "A Refutation of Russell's Stereotype," in *Whitehead: The Algebra of Metaphysics*, edited by Ronny Desmet and Michel Weber, 127–209 (Louvain-la-Neuve, Belgium: Les éditions Chromatika, 2010).

8    Actually, "being *cogredient* within a duration" is the expression Whitehead used in order to refer to the concrete intuition that ultimately gives rise to the abstract concept of "being at *rest* with respect to a reference frame."

9    For historical details on Whitehead's reception of, and alternative to, Einstein's general theory of relativity, see Ronny Desmet, "Putting Whitehead's Theory of Gravitation in Its Historical Context" *Logique & Analyse* 214 (2011): 287–315; and for an extensive comparison of Einstein and Whitehead, see Ronny Desmet, "Aesthetic Comparison of Einstein's and Whitehead's Theories of Gravity" *Process Studies* 45.1 (2016) and the book-length essay "Out of Season," *Process Studies Supplements* 23 (2016).

10    Whitehead distanced himself from Wittgenstein, too; see Chapter One of this anthology.

11    Thomas Nagel, *Mind & Cosmos: Why the Materialist Neo-Darwinian Conception of Nature Is Almost Certainly False*, Oxford University Press, 2012, and H. Allen Orr, "Awaiting a New Darwin" *New York Review of Books,* 7 February 2013.

12    For an extensive comparison of Poincaré and Whitehead, see Ronny Desmet's book-length essay "Poincaré and Whitehead on Intuition and Logic in Mathematics," *Process Studies Supplements* 22 (2016).

13    Harald Atmanspacher and Thomas Filk, "A Proposed Test of Temporal Nonlocality in Bistable Perception," *Journal of Mathematical Psychology* 54 (2010): 314–21.

# *1. Integral Philosophy: An Essay in Speculative Philosophy*

RONALD PRESTON PHIPPS

Alfred North Whitehead, while at Cambridge University, was co-author, and indeed principal author of the mathematics of, *Principia Mathematica*, which the esteemed German mathematician David Hilbert is said to have called the greatest axiomatic system in human history.

Approximately two decades later, at Harvard, Whitehead composed *Process and Reality*, which can be viewed as the greatest axiomatic system in the history of metaphysics. Whitehead called his metaphysics the "philosophy of process" or, as he preferred, "the philosophy of organism."

Now, one hundred years after the creation of Einstein's theories of relativity, physicists understand Einstein with more depth than Einstein did. That is the nature of axiomatic systems, whose full understanding requires centuries: first, to deduce major theorems; then, to allow those theories to interact with practice, observations, and questions leading to adventures of discovery. An infinity of logical consequences are deducible from the concepts, axioms, and theorems of any rich axiomatic system. This fundamental truth pertains to Whitehead's philosophy of process and organism.

There are vital and valuable roles to be played by commentary scholarship and exegetical analysis. Whitehead's philosophy, however, is an exercise of speculative philosophy. To be true to itself, process philosophy must be in continuous processes of creative development, creative application, and creative modification. The last thing Whitehead would want is for those inspired by his revolutionary philosophic system to reduce the system to mindless mantras. Whitehead's speculative philosophy was a philosophic opening to new vistas, the integration of disparate phenomena and disciplines of thought.

Whitehead's metaphysics describes a universe that is spatially infinite and temporally eternal, with neither boundaries nor borders, beginning nor end; it is pervasively organic, a universe all of whose finite domains are integrated and interactive with their spatio-temporal adjacent communities of being. All finite domains concurrently realize and frustrate the immense diversity of causal potentialities that dwell within each finite domain of the infinite cosmos.

Whitehead's metaphysics envisions an infinite, open, and integrated cosmos in perpetual flux. The basic premise of the universality of flux was captured by Heraclitis' premise "We can never step into the same river twice" and St. Benedict's statement "Always, we begin again." In Whitehead this enduring intuition assumes its most systematic and formal articulation and the most comprehensive investigation of its implications and consequences.

Whitehead's metaphysics, however, was disdained during the era of its creation. Indeed, all modes of metaphysics and "integrative philosophy" were out of favor during the dominance of analytic philosophy and its sub-domains of logical empiricism, logical positivism, and common language philosophy.

The aim of this essay is to revitalize "integral philosophy." As defined herein, integral philosophy advocates the integration of analytic and synthetic modes of thought. From such integration creativity, innovation, clarity, coherence, and discovery concurrently emerge.

## WHITEHEAD'S LETTER TO LEONARD AND LEONARD'S COMMENTS ON WHITEHEAD'S LETTER

In 1937 Whitehead wrote a letter to his personal assistant at Harvard, Henry S. Leonard. This letter was in response to the 1936 *Festschrift,*

*Philosophical Essays for Alfred North Whitehead,* composed by Whitehead's philosophy students at Harvard in celebration of the 75th anniversary of Whitehead's birth.

There was an aura of mystery surrounding the letter, written in response to Leonard's essay "Logical Positivism and Speculative Philosophy." The letter was ardently sought by Professor Victor Lowe, Whitehead's student and biographer, but not discovered. Fortunately, I was able to obtain a copy from Professor of Mathematics Henry S. Leonard, Jr.

Logical positivism and analytic philosophy, with their deep antipathy towards metaphysics, were the reigning rage, as Whitehead composed *Process and Reality.* Distinguished members of the Vienna Circle, including Rudolph Carnap and the great Viennese mathematician Karl Menger, visited Harvard during this period of fervent intellectual creativity. It was an era of sharply divergent schools of thought. Leonard was in a unique position to follow this evolving debate, and Leonard was the only one among Whitehead's distinguished students who shared Whitehead's deep passion for both mathematical logic and metaphysics.

The historic significance of Whitehead's letter to Leonard resides in the illumination it provides regarding several of his contemporaries and the drift of thought emerging as a counter current to the role of speculative philosophy and metaphysics as advocated by Whitehead.

With the invaluable assistance of Dr. Timothy Eastman and Dr. Carolyn Brown, we were able to celebrate the discovery of this letter of cardinal philosophic significance during a conference held in the Thomas Jefferson Memorial Room of the magnificent Library of Congress. Professor Henry S. Leonard, Jr., presented the letter to the library where it is now housed along with Leonard's formal comments.

Whitehead was widely revered for his humility, gentleness, and open mind. The letter to Leonard reveals a sharpness atypical of this sage. The letter begins:

> I want to concentrate on your work. It held me with intense interest. The lucidity and 'depth'—or rather 'width'—of your work are beyond praise. . . . Every mathematician and symbolic logician is, in his habit of thought, a logical positivist.

Yet to some of the expositions I find myself in violent opposition — especially to the very habit of dismissing questions as unmeaning i.e. unable to be expressed in existing symbolism. Wittgenstein annoys me intensely. He is a complete example of the saying

> I am Master of this College
> What I know not, is not knowledge.

Logical Positivism in this mood — its early mood — will produce a timid, shut in, unenterprising state of mind, engaged in the elaboration of details. I always test these general rules by trying to imagine the sterilizing effect of such a state of mind, if prevalent at any time in the last ten thousand years.

The handwritten letter proceeds to an unexpected appraisal of his student, colleague, and co-author Bertrand Russell; Ludwig Wittgenstein; and leading logical positivist Rudolf Carnap. Whitehead writes:

> The fact is that thought in the previous two centuries has been engaged in disengaging itself from the shackles of dogmatic divinity. Thus it unconsciously seeks new fetters . . . But I see no reason to believe that the stretch of Bertrand Russell's mind, or of Wittgenstein's mind, or of Carnap's mind, has attained the limits of insight or expression possible in the evolution of intelligent beings. They are bright boys, representative of a stage of rationalism, but nothing more.

The historic philosophic context of the letter was the ascendancy of logical empiricism, logical positivism, analytic philosophy, and the dominance of language, syntax, structuralism, and formalism in philosophy, literature, and other domains of intellectual inquiry. This trend emanated from the Vienna Circle. The constrictive trends of thought advanced by the logical positivists were an understandable reaction to muddled, pretentious and conceptually vague, and incoherent attempts at philosophizing. They were a reaction against obscurity, the lure of mysticism, and the proclivity towards subjectivist philosophies that dominated earlier centuries. Logical positivism led to a century of obsession with methodology over substance, and form over content. This resulted in a passive and subservient relation to science and a nihilistic, rather than creative, approach to the traditional concerns of philosophical inquiry. For Whitehead, philosophy had an essential role of challenging the underlying presuppositions of prevailing scientific dogmas, as well as finding modes of integrating truths of disparate disciplines.

Whitehead saw his metaphysics as a repudiation of "the Kantian doctrine of the objective world as a theoretical construct from purely subjective experience," and "the distrust of speculative philosophy" (PR xiii). Examination of presuppositions, analysis, logic, creativity, synthesis, speculation, intuition, and imagination are the tools of speculative philosophy.

Professor Leonard, after his Presidential Address to the American Philosophical Association, prepared to publish Whitehead's letter and his reflective comments. Leonard served for a decade as Whitehead's personal assistant when Whitehead's metaphysics was being formulated, debated, and refined. Leonard had pondered in subsequent decades upon the evolution of Whitehead's philosophy, which entailed modification, development, and synthesis of his previous work in mathematics, theoretical physics, and philosophy. That creative synthesis culminated in Whitehead's philosophic trilogy *Science and the Modern World*, his magnum opus *Process and Reality*, and *Adventures of Ideas*.

This philosophical trilogy in important ways both extends and supersedes Whitehead's trilogy (*Principles of Natural Knowledge*, *The Concept of Nature*, and *The Principles of Relativity*) in theoretical physics composed at Imperial College, London. Whitehead's philosophic trilogy represents, in ways not fully appreciated, an evolution, a modification, and a synthesis of this previous work, which is immanent within the philosophic trilogy. It is a trilogy representing a creative synthesis guided by the discipline of logical analysis in the context of formulating a historically novel philosophic vision of the universe.

After working with Gödel at the Institute for Advanced Studies in Princeton for two years, Leonard was invited as a visiting professor to Washington University. There Leonard presented his "Wide Language W," synthesizing propositional and modal logic. This synthesis reflects the underlying logic of *Process and Reality*, in whose metaphysics truth, falsity, possibility, necessity, and potentiality are fundamental parameters. It was during this period that Leonard composed his comments to Whitehead's letter.

Awareness of the incompleteness of process philosophy was keen among Whitehead's students who knew that the philosophy needed elucidation, modification, and exploration through (1) inductive processes of observation and accumulation of data, (2) deductive processes of logic, and (3) the exercise of human imagination envisioning new possibilities.

Leonard writes as follows:

> My indebtedness to Whitehead as an intellectual guide to my own philosophical position was sensed at the time this essay was written, but it was not as deeply felt nor as pronounced as later experience made it to become. The strongest influences operating upon me when I wrote this essay were those of (1) C.I. Lewis, as teacher, friend, colleague and author of *A Survey of Symbolic Logic* and *Mind and the World-Order*; (2) Rudolph Carnap, primarily as the author of *Der Logische Aufbau der Welt* and *Die Logische Syntax der Sprache*; and (3) Nelson Goodman, friend and fellow graduate student with whom I had for a long time the closest of intellectual relations. . . .

> As I reread the essay today, I find myself tempted to say that there is practically nothing in it that I would change, were I writing it today, except the whole thing! The style is too cocksure, the envisagement of problems too simple-minded. . . .

> The essay was, among other things, an effort on my part to explain what I then took logical positivism to be. But the experience of the next dozen years proved that people who were calling themselves logical positivists were generally not doing philosophy in the manner in which I had in this paper described the positivist as doing it. They were too busy ruling out whole areas of philosophy — such as metaphysics and ethics — as nonsense, instead of reconstituting those areas as valid fields of inquiry. Or they were . . . modeling metaphysics on epistemology. This latter tendency I already deplored . . .

> in 1948, I made a point of ceasing to call myself a logical positivist . . . This did not so much imply an abandonment of

doctrine set forth in the present paper as it implied an impatience with, first, what other people calling themselves positivists had averred and with, second, the speculative restraints characteristic of positivism. . . .

Also, their [the positivists'] inclination was one toward nominalism, and an extensional and a truth functional logic, whereas my bias . . . was and has become increasingly one toward a modified realism and a modal and an intentional logic. . . .

Philosophy aims at a unity, which is something more than can be achieved merely by a unifying of current scientific theories, and a speculative completeness, which also is something more than — and perhaps different from — what current scientific theories can provide. . . .

Finally, they saw philosophy as nothing more than analysis, an inquiry into the logic (the syntax) of (the language of) science; whereas I have seen then — and even to a greater extent see today — philosophy as an enterprise including speculative analysis, but as also including legitimate speculative inquiry which incorporates both a critique of current scientific speculations and a venturing beyond them.

## INTEGRATIVE PHILOSOPHY AND THE METAPHYSICAL MISSION

Integrative philosophy and the continual advance of the metaphysical mission are not fulfilled, as Russell put it "in one stroke." Integrative philosophy and metaphysics evolve from the confluence of the creation, refinement, and integration of concepts and generalizations which are themselves created from complex historical processes of cumulative inquiry, modification, processes of inductive and deductive reasoning, observation, and humanity's collective experience. In Whitehead's *Adventures of Ideas,* he describes with great eloquence the fact that knowledge is a collective, cumulative, and communal creation of our species that should tolerate no shallow and individual egotism. Individual genius exists only within communal genius.

The "three bright boys" to whom Whitehead refers in his letter represent three major subdivisions of the analytic philosophical tradition of the 20th century. Each school of thought historically served as a nihilistic and antagonistic force towards the metaphysical mission springing from ancient acts of wonder and curiosity.

Integrative philosophy seeks to avoid both (1) the vacuity of form without substance, and (2) the pretentiousness of philosophies offering substance without form. Synthesis undisciplined by logical analysis, precision, and deduction is lost in the vagueness of mysticism. Philosophy's mandate is to harmonize form and substance, and in that process enhance illumination of what is known while stretching human understanding through processes of discovery, innovation, and creativity to what is unknown.

Integrative modes of thought and analytic philosophy are two distinct phenomena. Of the former, Whitehead was a master, as manifested in his work as mathematician, theoretical physicist, and philosopher. The latter he repudiated, as did Leonard. For almost half a century analytic philosophy ignored Whitehead's metaphysics as irrelevant and meaningless. His metaphysics was too speculative, integrative, and boldly imaginative for that rationalistic tone of mind preached from philosophy's pulpits during the 20th century.

Revisionist attempts to confine within the box of analytic philosophy Whitehead's speculative metaphysics miss the philosophic impact of the historic letter of Whitehead to Leonard. Under the sway of the "three bright boys" and the philosophies they represented, philosophy became obsessed with the methodology of science and ignored philosophy's traditional role as gadfly, examiner, and challenger of the presuppositions underlying prevailing scientific worldviews. The metaphysical mission was subverted whenever it endeavored to climb to higher levels of generalization and wider modes of integration than allowed by prevailing presuppositions. Philosophy needs new, open, and integrative rubrics in accord with which it will creatively contribute to continuous adventures of discovery.

To understand the deeper meaning of Whitehead's letter to Leonard, it is necessary to consider Whitehead's understanding of the metaphysical mission. In its broadest framework, metaphysics consists of two interrelated domains: ontology and cosmology. Ontology seeks to understand the most fundamental constituents of reality, both concrete and abstract, inclusive of the real, the possible, and the potential. Cosmology concerns the most generic relationships among the constituents of reality. The most pervasive aspects of the cosmos consist in the community of being, since every finite actuality emerges from, dwells within, and perishes amid communities and interdependence. The universe is pervasively interrelated, including the orderliness by which the past creates the present and the present transitions to the future, both according to patterns that are the subject of empirical scientific inquiry. Causal orderliness is not merely the subject of scientific adventures of discovery but also the precondition upon which an effective teleology and the very eros of the universe emerge within the flux of being.

It is also the basis upon which the universe is rich in causal potentialities. A fundamental premise of Whitehead's ontology is that within any finite event, or aggregate of events, there resides a multiplicity of causal potentialities that are compatible for co-presence but incompatible for co-realization. It is the causal interaction within the wider community of events, the entity's environment, which influences which of those potentialities are realized and which are frustrated as the many become one and the one is added to the many. Whitehead describes in *Process and Reality* this inevitable realization/frustration duality as the possibilities "left unrealized within the womb of nature" (PR 17).

Integral philosophy is needed most fundamentally not because of epistemological demands, but because reality in all its domains is pervasively, inextricably integrated through processes of synthesis and immanence. Analysis without being infused and enriched by intuition, imagination, and synthesis is, in Macbeth's words, "full of sound and fury signifying nothing."

Concurrently, to reduce Whitehead's metaphysics to the mode of mysticism, such as those texts dominating the "metaphysics" sections in bookstores, is to (1) fail to understand the demands for clarity, precision, and coherency within the conceptual scheme, and (2) abandon the deductive and logical power of axiomatic systems. Whitehead, with an absence of timidity, revitalized metaphysics in a philosophic world that sought its annihilation. It is entirely appropriate to understand *Process and Reality*, as Whitehead did, as adding analysis and clarity to the ancient and valuable intuitions of Eastern philosophical traditions, including Hinduism and Daoism, with which he wove a modern philosophic tapestry.

Unlike Cartesian philosophy, which posits that to be is to be independent of everything else, "to be" for process philosophy is to be in community and to develop and evolve within a multiplicity of evolving relations. It is by processes of synthesis and integration that the Many become One. As Whitehead stated in *Essays in Science and Philosophy,* no finite entity can ever "shake off its essential connection with" its "infinite background" (77).

Integral philosophy is the antidote to the limitations of both mysticism and analytic philosophy. Integral philosophy is in the mode of creative scholarship and serves as the essential handmaiden to the metaphysical mission. Integral philosophy is the integration of analytic and synthetic modes of thought, spiced by intuition, imagination, and speculation in pursuit of the metaphysical mission. It is through the cultivation of integral philosophy that adventures of discovery and generalizations of insight emerge and continually advance.

As Whitehead stated in *The Aims of Education*: "The present holds within itself the complete sum of existence, backwards and forwards, that whole amplitude of time which is eternity" (14). No event can be separated from that community of events which is its past or that community of events which is its future. The community of being pervades the infinity of space and the eternity of time.

In the final analysis, Whitehead saw his philosophy as both integrating and revolutionizing the underlying philosophic presuppositions guiding all disciplines of inquiry and arenas of creation with insights demanding creative renewal and transformation. *Constant flow*

## WHITEHEAD'S METHOD OF EXTENSIVE ABSTRACTION, LEONARD'S CALCULUS OF INDIVIDUALS AND MEREOLOGY

The importance of part/whole relations was recognized in ancient Greece by Plato and in 7th-century India in the extensive studies of Dharmakirti. Polish mathematical logicians such as Alfred Tarski added a modern dimension.

Whitehead intended to write the projected fourth volume of *Principia Mathmatica* without Russell's participation. That volume was to address the philosophy and mathematics of geometry and was to present a revolutionary transformation of the ontology of geometry. This promised to be the first major change in the basic constituents of geometry since Euclid's great *Elements of Geometry*.

Rather than deriving the extended from the extensionless (lines from extensionless points, planes from lines, and volumes from aggregation of planes), Whitehead took regions with their extensiveness, their relations with other regions, and their essential metrics to be the fundamental constituents of geometry.

Indeed the regions within which the infinite flux of becoming occurs have spatial and temporal extensiveness, spatial and temporal relations, and spatial and temporal metrics. All else represents abstractions from the concrete and dynamic flux of events. Both spatial points and temporal instants are derivative entities from the spatio-temporal manifold.

As a mathematical logician, Whitehead engaged in the highest levels of abstraction, but in *Science in the Modern World* Whitehead introduced the concept of the fallacy of misplaced concreteness. The fundamental mandate of this fallacy is the need to understand the relation of the abstractions of thought to the underlying richness, complexity, and diversity of the concrete reality from which these abstractions are drawn.

*Process and Reality* contains Whitehead's last account of his method of extensive abstraction, which can be read as the projected transformation of the ontology of geometry that would have formed the ontological foundation of *Principia Mathematica* Vol. IV.

In *Process and Reality*, the two most fundamental categories of existence are eternal objects and events. Eternal objects provide the intrinsic modes of definiteness (characteristics/attributes) and the modes of relationship among concrete events or eternal objects.

Attributes and modes of definiteness create pure possibilities for events within the spatially infinite and temporally eternal manifold of events. Pure possible modes of definiteness do not determine whether or not, when or where, or among how many events they ingress. Modes of definiteness can have multiple ingressions within the flux of becoming and, thus, can recur. In contrast, events causally emerge from antecedent events and perish into the history of the cosmos. → *events come & go (just recur)*

The relationships among events within the spatio-temporal manifold include metric relations of spatial magnitudes, temporal magnitudes, shapes, sizes as well as generic geometric relations of covering, over-lapping and being separated from, and temporal relations of "before," "after," and being separated by this or that temporal magnitude.

The most fundamental feature of the manifold of events is exten-siveness, which is divisible by an infinity and continuity of magnitudes. All events have unique spatio-temporal loci within the extensive con-tinuum of events. While the spatio-temporal manifold is continuously divisible, it is, for ontological reasons, discretely divided by atomic occa-sions, quantum events, or "Puffs of Becoming." [1]

Space and time are constituted by an infinite number of possible modes of spatial magnitudes, shapes, and relations. Space is neither created nor destroyed. Nor does space expand, contract, or curve under the influence of events. Expansion, contraction, and contortion neither create, nor destroy, nor change space. Expansion, contraction, or con-tortion of any finite domain of events presume and presuppose space as the realm of infinite spatial magnitudes, shapes and relations which bind events in unique locations and complex relations within the manifold of events. The same pertains to the infinite realm of temporal magnitudes and relations binding an infinity of events in temporal relations.

To conceive of space or time as contingently changing, for example, as curving or dilating, is to misunderstand, as Einstein's General Theory of Relativity does, the ontology of space and time. Space does not curve. Temporal sequences of events may form curved paths or trajec-tories under the causal influence of fields of events defining the environ-ment which these temporal sequences transverse, and with which they interact. Similarly time does not dilate. Sequences of events exhibiting periodicity, such as clocks, may dilate as the temporal magnitudes of the event-members within sequences of periodicity differ.

In this Whitehead repudiates the world as a subjectivist construct, as postulated by Kant, and as subsequently manifested in the logical positivists' view of time and in the philosophic presuppositions under-lying both Einstein's Special and General Theories of Relativity.

Euclid's *Elements* and the subsequent tradition of geometry for two millennia posited extensionless points from which extension was to be derived. The obvious philosophic and conceptual contradiction is the ontological inability to create extended magnitudes and the metrics characterizing the reality of the cosmos by the aggregation of an infinity of extensionless and mythical entities. An infinity of zeros is zero.

*Process and Reality* begins its ontology of geometry with the extended. Points (lines, planes) are derivative entities consisting of infinite sequences of regions defined in respect to relations of "cover-ing" and "overlapping" that converge to a given "point" ("line," "plane") in respect to space, and an "instant" in respect to time. Whitehead, therefore, inverted the relation of the extended and the extensionless.

Leonard's 1930 doctoral dissertation at Harvard established a formal theory of the part/whole relation among extended entities. Leonard's dissertation evolved into "The Calculus of Individuals and Its Uses," published in 1940 by Henry S. Leonard and his Harvard colleague Nelson Goodman. Leonard was the principle author of the mathemat-ical logic of the Calculus of Individuals.

For his contribution, Leonard is regarded as the father of mereology, a mathematical field that explores the relations of parts and the wholes of which the former are ontological constituents. During the 1970's there was a revival of Mereology initiated among European mathematicians and computer scientists.

Mereology is a sister to Whitehead's Method of Extensive Abstraction. It allows us to overcome the mythical geometrical extensionless entities which were posited over two millennia ago, and from which extension was to be derived. The Method of Extensive Abstraction and Leonard's Calculus of Individuals and Mereology derived therefrom allow us to create a systematic understanding of the universe as constituted by spa-tially and temporally extended events embedded within relations of extensiveness and characterized by essential metrics which delineate the temporal and spatial magnitudes of the perpetual processes of becoming within a boundless and borderless manifold of events.

It is interesting to note that the great Austrian mathematician Karl Menger's famous Theory of Sponges has its starting point in extended three-dimensional volumes. Analogously to Whitehead's derivation of geometric points, lines, and planes from infinite converging series of extended regions, Menger investigates the results of using formulae to extract an infinite series of sub-volumes from Menger volumes similar to ancient Chinese carved ivory spheres. Menger's seminal work has great scientific relevance to fractals and the patterns found among a wide range of natural phenomena inclusive of biological, geological, and ecological phenomena.

We may safely anticipate that there will be many future examples of the philosophic, mathematical, and scientific value of taking as one's starting point the extended and relations among extended entities. Menger's mathematical discoveries enjoy an affinity to Whitehead's transformation of the ontology of geometry.

Using the basic concepts and premises of Whitehead's ontology of geometry and his Method of Extensive Abstraction we can define concepts fundamental to the cosmology of *Process and Reality*.

All events occur within "continuous" spatio-temporal regions. This includes (1) huge and dynamic cosmic epochs, as well as both (2) "atomic occasions" or elementary events, and (3) organic aggregates or nexuses of events. Here are some definitions:

A region $R_x$ is continuous if, and only if, for any regions $R_y$ and $R_z$ where $R_x$ covers $R_y$ and $R_z$, and the fusion of $R_y$ and $R_z$ equals $R_x$, $R_y$ and $R_z$ are contiguous.

A spatial domain $D_x$ is infinite if, and only if, for any spatial magnitude M, there is a spatial domain $D_y$ such that M characterizes $D_y$ and $D_x$ covers $D_y$.

A spatial domain $D_x$ is finite if, and only if, for some spatial magnitude M there is a spatial domain $D_y$ such that M characterizes $D_y$ and $D_y$ covers $D_x$. (For Whitehead to be finite is to have a beyond.)

The manifold of events (the cosmos) is eternal in two directions if for any finite event $E_y$ within the manifold of events and for any temporal magnitude M, there are events $E_x$ and $E_z$ such that $E_x$ is before $E_y$ and $E_y$ is before $E_z$ and ($E_x$ and $E_y$) and ($E_y$ and $E_z$) are separated by temporal magnitude M.

For the definitions cited above there are many hybrids that can be developed. This ontology of geometry leads to a cosmos or universe that is spatially infinite, temporally eternal, and continuous.

These and similar definitions and axioms are essential to the creative development of Whitehead's cosmology. They include the metric dimensions of the spatio-temporal continuum whose constituent events emerge and perish within the framework of matrix characterizations.

Space is ontologically conceived as the infinite totality of all possible spatial shapes, magnitudes, and relations, and time as the infinite totality of all possible temporal magnitudes and relations that may characterize the infinite and eternal manifold of events. The evolution of aggregates of events, including cosmic epochs, ontologically presumes these infinite realms of spatial and temporal relationships.

## WHITEHEAD'S COSMOLOGY AND BIG BANG MYTHOLOGY

The need for fundamental changes in the philosophical presuppositions underlying physics is not a transient but a permanent imperative. Leonard made it clear that this was Whitehead's view. Whitehead's cosmology offers resonance with many developments in 20th century theoretical physics, including relativity theory and quantum theory, but in the final analysis it transforms underlying presuppositions in such a manner as to promise revolutionary change.

Whitehead perceived in ancient Eastern modes of thought both a challenge to prevailing Western presuppositions and a proclivity to envision greater qualitative diversity, enormity of scale, and more pervasive organic interdependencies. Whitehead's perceived revolutionary changes were both called for, and dependent upon, a deeper synthesis of the philosophic predilection of the ancient intuitions of Eastern and modern Western modes of thought.

Whitehead's concerns with the inadequacies of the underlying presuppositions of 20th century physics include opposition to the lingering effects of the presumption of atomism, reductionism, subjectivism, finitude, the rigidity of a universe governed by constants rather than orderliness, the independence of entities, the bifurcation of nature, and the tendency to see things in terms of strict dichotomies rather than polar contrasts. All of the above philosophic tendencies lead, as Whitehead expressed it, to a closed state of mind.

When these philosophic tendencies lead to contradictions, the "cure" tends to be in *deus ex machina* solutions that lead to journeys into the bizarre; not the coherent and logical consistency which the mathematical logician sought to impart to his evolving metaphysical cosmology. The intuitive reach to novel vistas and broader integration must be disciplined by coherency and consistency.

A crucial imperative of Whitehead's cosmology is the creation of a conceptual and axiomatic scheme in terms of which the phenomena studied in both the "hard sciences" of physics, chemistry, astronomy, and biology, and the soft sciences of psychology, sociology, history, and the arts can be reconciled and integrated as representing abstraction from the complexity of reality. To achieve this integration there must be a clearer integration of causal orderliness and the presence of teleology in the universe, the physical and the mental, truth and beauty, fact and imagination. Without causal orderliness neither would teleology or high levels of mentality emerge during the evolution of the universe in its creative advance. It is the teleology to create, innovate, discover, and understand that impels science itself to participate in the creative advance of knowledge.

Big Bang mythology offers a vision of a singularity totally incoherent with all phenomena experienced individually or communally by the human species. The Big Bang theory posits a singularity with neither temporal nor spatial extension, structure, character, or explicable dynamics; a singularity devoid of origin, history, and environment, and even preexisting the emergence of either time or space; a singularity within which all being, matter, and energy are condensed; a singularity of infinite density whose explosion leads to a universe of infinitesimal density! The universe is conceived as emerging from the penultimate community without diversity, a singularity which would devolve into absolute insularity. This is a rather immodest conception for which no analogy to experience exists within known phenomena and processes. And yet in the 20th century this extravagant conception entered a realm of unassailable dogma, a dogma inexorably linked to particle physics.

Einsteinian reflections involving gravitational, electromagnetic, and strong and weak nuclear forces, led to predictions that the Big Bang would slow down, reverse course, and lead to the Big Crunch. Contrary to expectations, the astronomical data employing standard interpretations led to the conclusion that the finite universe not only continues to expand but is accelerating, and in a manner such that the unexpected acceleration is itself accelerating. To the rescue came various *deus ex machina* concoctions, including the revival of Einstein's cosmological constant, dark matter, and dark energy all based upon the perceived need to find a repulsive force which with increasing distance, and contrary to all known forces, increased in intensity and strength.

The false presumption of a temporally and spatially finite universe underlies a plethora of *deus ex machina* hypotheses, incoherencies, and exercises in the bizarre. The restoration of coherence can be achieved only by the philosophical presupposition of an infinite, eternal, open, integrated, and organic cosmos.

The development of Whitehead's cosmology leads to an infinite, open, and integrated universe evolving within an infinite sea of cosmic epochs emerging, evolving, and perishing in accord with their internal dynamics and their causal interaction with adjacent spatio-temporal epochs.

Each cosmic epoch is *qualitatively* delineated in an open, non-static universe, and not merely by the changing borders and boundaries of its spatial and temporal extensiveness. To say that the universe is open is to say that it is open to infinite qualitative variety beyond that diversity characteristic of our cosmic epoch or any other cosmic epoch. Ontologically the universe is open to multiple forms of energy beyond those that characterize any given cosmic epoch.

The delineating of cosmic epochs is not quantitative but qualitative. To understand the breadth of cosmic diversity inherent in Whitehead's mature cosmology, we view forms of energy that are defined by the mathematical, logical concept of Quantities. Quantities are infinite sets of qualities that can be ordered by "larger than" relations and for which additive and cumulative relations pertain.

Different cosmic epochs are distinguished by the sets of Quantities (forms of energy) that "cover" them. The cosmology of an infinite sea of cosmic epochs allows a diversity of sets of Quantities or forms of energy to cover qualitatively diverse cosmic epochs that arise and perish within the flux of events. Each cosmic epoch is spatially and temporally finite. This cosmological theory is compatible with an immense depth and breadth of internal and external qualitative diversity.

This theory can incorporate Big Bang phenomena of expansion, acceleration, de-acceleration, rhythmic change, and cessation within the infinite sea of cosmic epochs. It can do so without any hypothesis of a universe erupting from a state of infinite density, experiencing increasing acceleration of expansion, or devolving into a nihilistic vision of zero density. The cosmology of an infinite and eternal cosmos with neither borders nor boundaries is a vision of perpetual and pervasive cosmic creativity.

The universe is integrated in that all events and all aggregates of events, including all cosmic epochs, emerge, evolve, and perish, and exist within larger communities of events bound together in causal interactions, exchanges in forms of energy, and the concurrent realization or frustration of causal potentials inherent within and among cosmic epochs.

There have been critical analyses by Geoffrey Burbidge, Timothy Eastman, Eric Lerner, and Hilton Ratcliffe of the "evidence" offered for the Big Bang. Among other things, these critiques concern how we interpret data of light transmission through vast and unobserved regions of space-time, including dynamic plasmas, and complex and energetic fields. There has also been a shaking of the presumption of cosmic finitude. The Russian physicist Andrei Linde at Stanford has led this critique and offered a concept of multiuniverses that enjoys a constructive relationship to Whitehead's preliminary concept of cosmic epochs. These multiuniverses are envisioned as possibly accessible through black holes, time warps, and other exotic constructs. But they are conceived as essentially independent, different, and vast universes.

The concept we present is of a universe that is infinite, open, and integrated. To be pervasively integrated means that every finite region of the infinite and eternal manifold of events holds within itself a multiplicity of causal potentials whose realization or frustration reflects the internal dynamics of that region and the causal interaction of that region with its neighbors. Each cosmic epoch emerges and holds within itself immense complexity and causal potentialities to rhythmically or cyclically expand and contract, accelerate and decelerate, and emerge and perish. For this reason the Big Bang theory is subsumable within our model.

The philosophy of organism maintains that the fundamental types of processes found in large cosmic domains are also found within smaller domains of events. For example, the self-replication of living organisms depends upon processes of multiplication, differentiation, and integration. As living organisms develop the constituents of their embryos enter sub-environments whose physical, chemical, and biological characteristics differ leading to the realization and frustration of the causal potentials that are defined by the organism's genome and common to all of those constituents. Such processes result in differentiation and integration and depend not merely upon the internal dynamic but the interaction of the evolving organism with its environment. This is analogous to the internal processes of the organisms with the organism's interaction (absorption and expulsion of energy) with its environment. This is exactly what happens in the development and evolution of each cosmic epoch evolving within an infinite sea of cosmic epochs. This organic metaphor between living organisms and cosmological processes conforms to the intuitions of research physicians more than the prevailing dogmas that have captured contemporary astrophysicists.

## WHITEHEAD'S COSMOLOGY AND PARTICLE PHYSICS

Whitehead's ontology is based upon the primacy of events over particles. Particles, or enduring objects, are derivative entities or sequences of events which enjoy spatio-temporal continuity, contiguity among successor constituents, and the ingression of a set of attributes definitive of the character of the sequence and the type of particle generated and creatively sustained through their ingression. The ultimate constituents are events perpetually emerging ("Puffs of Becoming") and perpetually perishing within the flux of being. Events attain ontological primacy over particles, meaning that relative permanence or temporal endurance is not an intrinsic quality of particles but an achievement of causal interactions. Every atomic or elementary event contains, within its multiplicity of causal potentialities, potentialities for cessation and endurance, for repetition, for proliferation, and for novel creation. This ontological perspective is consistent with, if not derivative from, the most fundamental insight of Einstein's Theory of Relativity, viz., the equivalence of mass and energy, which in Whitehead's more open and expansive ontology and cosmology is expressed as the generic transformation of forms of energy one into another.

Permanence or semi-permanence reflects spatio-temporal contiguity and qualitative continuity among the events constituting sequences of events and aggregates of sequences, whether those aggregated sequences are simple or complex, small or large.

This ontological primacy also governs light rays (photons) and other sequences of events which constitute wave-like paths of various amplitudes and frequency. Fields of events, including gravitational and electromagnetic fields, are all subject to the primacy of events, i.e., the causal potentials to perpetuate, alter, or terminate the sequence, ray, or field.

The philosophy of organism, contrary to Descartes' philosophy in which to exist is to be independent of all other entities, posits that all finite entities exist within a community of being. Each event or process causally emerges from an antecedent community of events (its causal past), dwells within a contemporary community of events (its contemporary world), and causally contributes to a subsequent community of events (its causal future). The fundamental ontological principle is the universality of the community of being. In Whitehead's metaphysics the community of being prevails over the insularity of being.

The philosophy of process, in contradistinction to the presuppositions of both ancient Greek atomism and contemporary reductionism, is amenable to infinite qualitative diversity and variety within the realms of micro, macro, simple, and complex phenomena. It is essential to affirm the validity and value of both atomism and reductionism but Whitehead's ontology goes beyond their limitations.

Atomism and reductionism have enormous heuristic power in explaining, through physics, chemistry, and biology, a great deal of the qualitative diversity manifested in the world by means of a lesser qualitative diversity among constituents, which are subject to aggregation through the forces of nature and the infinite diversity of geometric forms. The essential premise of reductionism is that as we progressively examine the constituents of aggregates and the constituents of those constituents, qualitative diversity diminishes or is explicable merely through geometric forms that infuse the macro world with qualitative diversity.

The variety of generic causal potentialities within the atomic occasions at the micro level of existence include the causal potentials to continue, duplicate, and generate qualitatively different atomic occasions that may be parts of qualitatively different types of enduring objects or elementary particles, or to cease. The world of atomic occasions is a realm of enormous diversity. The large and powerful particle colliders in CERN, Brookhaven, Fermi Lab, Stanford, etc., are conceived as "Cauldrons of Dissociation" by which hidden constituents are revealed. In Whitehead's ontology we may view such colliders, as well as stars, plasmas, etc., as "Cauldrons of Creation" that through intense and dynamic forces provoke both (1) the dissociation of condensed and bonded constituents and (2) the creation and emergence of new types of atomic occasions and elementary particles.

The aversion to, and presumption of, the collapse of qualitative diversity is deeply entrenched in the presuppositions guiding physics, especially particle physics. Maria von Kryzwoblocki described to me the sense of intellectual distress among physicists as the number of types of elementary particles approached 2000. Such a proliferation of qualitative diversity (even one of such a small magnitude) is deeply disturbing and intellectually intolerable. The Standard Model, with its hypothesis of quarks, gluons, leptons, etc., with various spins, colors, flavors, etc., arose in service of the presumption of reductionism. Aspiring to supersede the Standard Model, string theory takes the impulses of reductionism even further. All concrete entities are conceived as strings in thirteen or more "dimensions," which differ one from another in virtue of their "vibrations." Thus, all qualitative diversity is supposed to be reducible to geometry, that is, the specific geometric mode of vibration (involving amplitudes and frequencies) of "strings" which otherwise are intrinsically homogeneous. Though vibratory and rhythmic behavior is essential to understanding the internal and external dynamics of events and their aggregates, in process ontology qualitative diversity is not reducible to mere geometric diversity.

The generic nature of the causal orderliness of the cosmos can be expressed in the formalism of mathematical logic. I express this formalism elsewhere.[2] We also formally deduce from the causal orderliness of the universe that the ingression of any attribute A in any event $E_x$ creates co-present causal potentialities $P^1$ and $P^2$ to influence the attributes and relations of subsequent events which, though compatible for co-presence, are incompatible for co-realization. The co-presence of causal potentialities in events that are incompatible for co-realization pervades all events whether micro atomic events or aggregates (whether small, medium, or large, and whether simple or intricate). The phenomena studied by astrophysics, chemistry, biology, psychology, and sociology jointly and separately illustrate this profound and pervasive ontological theorem.

Thus, the co-presence of causal potentialities incompatible for co-realization derives from the causal orderliness that pervades the cosmos. The ontological conjecture is an antidote to the reductionism whose elucidating powers are fully recognized but whose limitations are concurrently affirmed.

The ontological conjecture seeks to harmonize the causal orderliness of the universe with the infinite qualitative diversity residing at the core of existence. If field equations govern the generation of events from their causal pasts, and these field equations guiding the processes of confluence and creation are subject to infinite solutions, then from the very orderliness by which the qualities and relations of the confluence of antecedent events (the causal past) causally influence the intrinsic qualities and relations of the atomic events resulting from this confluence, the universe is both causally ordered and amenable to infinite qualitative diversity.

Particle physicists from Stanford and Fermi have discussed the ontological conjecture. It represents a fundamental change in perspective and challenge to the underlying presuppositions for the strategic investigation of the elementary constituents of reality. Whitehead's ontology, if creatively developed, offers a transition from the reign of sameness to the reign of diversity. Arbitrary restrictions and limitations on diversity are transformable to a synthesis and integration of openness, diversity, and orderliness. Dissociation is not the fundamental aspect to the ontological role of fields. Integration, synthesis, and creation are fundamental.

## OTHER DOMAINS OF INQUIRY AND CONCLUSION

There are many other areas of physics and science where Whitehead's metaphysics promises revolutionary transformations. Such areas include relativity theory, quantum theory, the Higgs field and the ontology of forces, and emergent attributes, both physical and mental. The confines of this paper limit detailed discussion of these topics, as follows.

1. *Relativity.* In Whitehead's mature ontology the concept of "congruence" is seen as an objective ontological relation rather than a subjective epistemological construct. This changes the interpretation of length contraction and time dilatation to apparent phenomena as Lorentz viewed his transformation equations. It also leads to ontological contradictions in Einstein's fundamental premise of the constancy of the velocity of light with respect to all objects in relative motion one to another. It supports Gödel's deduction that the Theory of General Relativity implies "A World without Time." The velocity of light is neither constant nor the maximum velocity for all cosmic epochs throughout the infinity of space and the eternity of time. In an open cosmology of the infinite, there is no necessity for any velocity to be a maximum velocity aside from specific cosmic epochs or special sub-regions of a given cosmic epoch. Light rays, like all enduring objects and sequences of events, have the causal potential to be bent in various trajectories, accelerated or decelerated as they traverse various mediums, plasmas, and fields. The transformation of forms of energy one into another cannot be fully explained by $E = mc^2$.

2. *Quantum Theory.* Whitehead significantly changes Heisenberg's

epistemological and subjective basis for quantum phenomena to an ontological and objective base. While the spatio-temporal continuum is discretely divided as atomic events emerge and perish, that discreteness within Whitehead's open ontology is not confined to integral multiples of Planck's constant. The transmission of causal influence is finite, variable, and not constricted by the velocity of light.

3. *Emergent attributes.* In the course of the evolution of the universe complex, refined and dynamic modes of aggregation of events occur which give rise to emergent attributes that characterize the whole aggregate (or specific domains within it) but do not characterize all its parts. Emergent attributes cover both physical and mental attributes and causally arise from both the physical and/or mental attributes of the antecedent events that generate the aggregates. Some of the most significant attributes, including consciousness, teleology, creativity, imagination, and intuition, belong to the realm of emergent attributes. The causal and bi-directional interplay of physical and mental attributes is crucial to the presence in complex living organisms, sociological communities, ecological systems, and complex physical structures of emergent attributes within the evolution and creative advance of the universe. From the ingression of complex emergent attributes into aggregates that are highly structured, integrated, refined, and dynamic powerful causal potentialities of creativity, innovation, discovery, deduction, synthesis, imagination, refined feeling, and teleology arise. The dynamics of emergent attributes represent a fundamental dimension of the ontology of organism that requires further philosophic articulation and scientific inquiry.

The creative development of Whitehead's philosophy of process and organism has enormous transformative potential for the above domains of inquiry. A detailed discussion of these topics can be found in the publications of this chapter's author, listed below.

The philosophy of process and organism holds the promise, if creatively developed, applied, and modified, to revolutionize and integrate our understanding of an infinite and organic universe in its perpetual processes of becoming.

## NOTES

1   "'Actual entities'. . . . differ among themselves: God is an actual entity, and so is the most trivial puff of existence in far-off empty space" (**PR** 18).

2   "The Philosophy of an Infinite, Open and Integrated Universe." In *Applied Process Thought II*, eds. M. Dibben & R. Newton. Frankfurt: Ontos Verlag, 2009.

## REFERENCES

Leonard, Henry S. "Logical Positivism and Speculative Philosophy." In *Philosophical Essays for Alfred North Whitehead*. New York: Russell and Russell, 1936.

Leonard, Henry S. and Nelson Goodman. "The Calculus of Individuals and Its Uses." *The Journal of Symbolic Logic* 5.2 (1940): 45–55.

Quiring, John. "Whitehead and the God Particle (The Higgs) with Ron Phipps." *Process Perspectives* (2013).

Phipps, Ronald Preston. "The Background and Historic Significance of Alfred North Whitehead's Letter to his Personal Assistant Henry S. Leonard: The Relation between Science and Philosophy." *Process Studies Supplements* 17 (2011).

Phipps, Ronald Preston. "The Philosophy of an Infinite, Open and Integrated Universe." In *Applied Process Thought II*, eds. M. Dibben & R. Newton. Frankfurt: Ontos Verlag, 2009.

# 2. Reflections on Intuition, Physics, and Speculative Philosophy

TIMOTHY E. EASTMAN

John Cobb, a prophet of modern times, describes how there is little hope that we will find an adequate and "healthy response to our global crisis in secularism and the institutions it has generated . . . modern philosophy, modern science, the modern university, and economism. I see them all as cutting themselves off from the broader stream of human experience and the wisdom it has generated" (Cobb 2014, 297). Similarly, via our consumerist culture, there are many examples in which materialist orthodoxies and secularism tend to dominate, encouraging cultural mores and worldviews that undermine any sense of self-transcending meaning and value, much less any consideration of the reality and scope of intuition.

In my undergraduate philosophy of science class (mid-1960s, University of Minnesota), Herbert Feigl (in 1924, a founder of the Vienna Circle of Logical Positivism) summarized how there are two primary types of philosophy, "philosophies of the something more, and philosophies of the nothing but." To illustrate this, following a lecture emphasizing the "overwhelming" evidence in support of the "nothing but" approach, the very next lecture began with a vignette recounting how he was still reveling in the sheer joy of having listened the previous night to Beethoven's 6th symphony, an exquisite musical evocation of the Vienna woods where he often walked in his younger years. How ironic to hear such mindful expression of great feeling and affirmation of value-talk by a founder of the Vienna circle, known for its "nothing but" philosophers!

A reductionist strategy argues that "aesthetic, moral, and mental properties supervene upon physical properties" (i.e., that there is some unique one-to-one mapping from macroscopic properties to microscopic properties).[1] Such supervenience roughly updates Feigl's model of psycho-physiological parallelism, albeit without the appearance of some magical alignment of events across two very different categories, which subsequently undermined the coherence of Feigl's and, indeed, Descartes' more extreme mind-body dualism. The new reductionisms that most frequently deploy supervenience tend to presuppose the following:

(1) that the microphysical (ultimately, the underlying physics) supports the kind of substance-talk or "substance-predicate" language most often ascribed to ultimate particles,

(2) that, ultimately, causal relations are exclusively materialistic "efficient causal" relations and, finally,

(3) that all "meaningful" propositions can be reduced to "nothing but" scientific propositions; i.e., "scientism" or a "scientistic metaphysics."

Although such presuppositions can be helpful for focused questions within certain limited domains of investigation, important progress in the conceptual foundations of biophysics, complexity and emergence (Clayton 2011) have reinvigorated a wealth of arguments and specific examples of why all three of these key presuppositions fail overall in light of the best of contemporary science and philosophy.

## ON SUBSTANCE-TALK

Rescher has given a concise summary of Johanna Seibt's argument that "the idea of a substantial object as standardly conceived in the philosophical literature is logically incoherent" (Rescher 1996, 64). Another critique of "substance" is provided by Lorenz Puntel:

> The root problem [of substance-talk] can now be formulated as the problem posed by predication on the level of first-order

predicate language . . . in which . . . the value of the bound variable x is simply and absolutely presupposed. The vital problem
is then this: such a presupposed entity is not intelligible . . . If
all the attributes (properties and relations) and whatever else
this presupposed entity may be involved in or within are taken
away . . . nothing determinate remains . . . Such an entity is
not intelligible and therefore must be rejected . . . Attempts to
rescue the idea of such a subject have been unsuccessful . . . [due
to] infinite regress [and other problems]. (2008, 194)

Another rescue strategy has been the common presumption of
"grammar as a guide to ontology" (McHenry 2015, 33), which is argu-
ably coherent within the confines of Western languages [Greek, Latin,
French, English being from the same linguistic family and sharing the
same substance-predicate structure] but then seriously fails when applied
to some Asian languages such as Chinese, in which the notion of process
is more fundamental.[2]

As Peter Hare has argued, Whitehead's "movement to the recognition of being as process rather than substance was as profound a
development in metaphysics as his work with Russell had been in math
and logic" (Hare 1995, 909).[3] As pointed out by Leemon McHenry
(2015, 61), it was no accident that Whitehead, having studied in the
1880s at Cambridge with prime expositors of Maxwell's new electromagnetic field theory (1861–62), namely W. D. Niven, in addition to
Edward Routh and J. J. Thomson (Desmet 2010, Ch. III)—and with
his exceptional rational intuition—eventually came to this insight early
in the 20th century. Indeed, Maxwell's new field theory was the focus
of Whitehead's fellowship dissertation at Cambridge.

More recently, in a detailed analysis of the concepts of "substance"
versus "event," Leemon McHenry demonstrates "not only that [an event
ontology] makes sense but that modern physics requires it" (2015, 55).
Such a consequence is not surprising given that contemporary physics
is fundamentally built on field theory (quantum field theory, relativistic field theories, etc.). What may be surprising, however, is that it has
taken more than a century to (just partly) shake off the presumption of
substance-talk as more fundamental than process and relations, whereby
substance-talk continues its hold through our default "world view of

perceptual objects" and subject-predicate linguistic traps. On this issue, as with many other fundamental interpretive issues in modern physics "There is . . . no escape from philosophy in the same way that there is no escape from some conceptual framework [or other] by which we interpret our experience" (McHenry 2015, 213). One scientist who tried to escape philosophy was the famous quantum theorist Richard Feynman. At one point, he called attention to how high-energy physics could be equally well called "wave physics" or "particle physics." Although the latter term has stuck, the choice is inevitably, in part, a philosophical question — indeed, such questions are part of the growing and active field of philosophy of physics. For a recent example of a noted philosopher-physicist collaboration that calls attention to both the importance of philosophy and science, as well as to becoming (process) and being, see Unger and Smolin (2013).

## ON RELATIONS AS EXCLUSIVELY "EFFICIENT CAUSAL"

As shown extensively by recent works in ecology, complexity, and emergence, essentially all natural systems are open and dynamic, rather than closed and static. "The creative advance manifests itself in two distinct ways: 1) internally, as an *individuating* process of self-integration, and 2) externally, as an *efficient* causal process whereby *individuated*, self-integrating structures determine or condition the self-integrating activity of successive individuals or structures" (Rose 2005, 12). The "individuating process of self-integration," which transcends efficient causal description, is revealed in at least four ways:

### (1ST WAY) PREHENSION

This concept is exhibited in detailed analyses of human experiencing, as in the radical empiricism of the psychologist and philosopher William James, as well as many philosophers in the phenomenology tradition. In his concept of "prehension," such self-integrative experience was hypothesized by Whitehead as an appropriate model for "experiencing" more generally.[4] Whitehead's key creative intuition about prehension is highlighted in Isabelle Stenger's classic monograph, *Thinking with Whitehead* (2011), which invites us all to wonder "what happened to Whitehead in

April, 1925" upon his key intuitive moment. Hartshorne (1979) has heralded prehension as the heart of "an unprecedented theory of creativity.

## (2ND WAY) EMERGENCE

Building on the works of Stuart Kauffman (2016), among others, Mirko di Bernardo recently summarized some key themes in complex system dynamics:

> In a complex system . . . the generation of new information is intrinsic to the dynamics of the process . . . the system continuously redefines the space of the alternatives . . . the centuries-old contraposition between *reductionism* and *naïve holism* today has been definitively surpassed by a new systemic vision according to which life appears as a transactional *phenomenon* resulting from a series of continuous arrangements constituting and unexpectedly modifying the parts of the game itself . . . [and] meaning reveals itself with time. (2015)

## (3RD WAY) THE "REAL" AS BOTH THE ACTUAL AND THE POSSIBLE

Recent research has revealed close mappings of phases of the quantum history of events (comprising all actualities at multiple levels) to Whitehead's analysis of actual entities. Such is found both in detailed analyses of contemporary relativistic quantum field theory — as in Verlinde's emergent gravity model (Bettinger 2015) — as well as in recent advances in solving the quantum measurement problem in which explicit use is made of both efficient causal and "logical" causal structures (Epperson and Zafiris 2013). Using well-tested elements that are core to quantum physics, Michael Epperson had demonstrated an exceptionally close mapping between Whitehead's analysis of actual entities and the fundamental structure of quantum events (Epperson 2004, 2012). The newest results by Epperson and Zafiris leverage new mathematical tools of category theory in such a way that I propose substantially advances our understanding of the quantum measurement issue, viz. (1) by highlighting the importance of the distinction of (ontological) possibility and actuality within the "real"; (2) by emphasizing the logical distinction, arising from category theoretic analyses of fundamental topological constraints, of local (Boolean algebra) and the (nonBoolean) global; (3) through recognizing histories of quantum events as truly fundamental; (4) by affirming the need for a logical causality[5] (demonstrating that the "efficient causal" description alone is intrinsically incomplete for science); and (5) by showing how the analysis of "possibility" requires careful philosophical examination, and explicitly adds to those critical factors (e.g., causality, temporality, values/choice) that are presupposed by both scientific practice and normal human experience.

## (4TH WAY) ON PERSPECTIVE AND ONTOLOGIZED "DECISION"

An application of the Conway and Kochen Free-Will Theorem — reflecting the self-integrative character of fundamental events — suggests that the presupposed possibility of "choice" by an experimenter at the classical level necessarily maps to some element of "choice" at the quantum level, however rudimentary, thereby suggesting that some elementary selection among alternative possibility ("decision") is inevitably involved in quantum events.[6]

Such "selection" need not be associated with a panpsychist view; however, it is consistent with what David Griffin (2016) has described as "pan-experientialism" in which the fundamental succession of events involves both prehension (active integration of past fields, particles, and other input into a unity) and concrescence (particular actualization of that new unity, the "actual entity").[7] Another way to understand this point, without depending on Whitehead's more technical vocabulary, is in terms of perspective(s) on given environment(s) as inevitable for any finite, contingent physical entity (see Ross 1983).

In addition to the above features of the Free-Will Theorem, Conway and Kochen also demonstrate "that there can be no correct relativistic deterministic theory of nature" (2009, 231). Further, that "combatibilism" is a "now unnecessary attempt to allow for human free will in a deterministic world" (230).

## ON SCIENTISM

With regard to the third key presupposition of "scientism" (indeed, to many framings of "physicalism"), I now address the common presumption that all "meaningful" propositions can be reduced to scientific propositions; i.e., a "scientistic metaphysics." In addition to the inevitable philosophical issues raised above, some variant of "absolute determinism" is often presupposed in such "scientistic" narratives, most often referred to simply as "determinism." Of course, as a principle of order, its opposite "indeterminism" can be similarly incoherent. Nevertheless, Shields (2016) has now concisely demonstrated the incoherence of strong determinism by using only the conservative assumption, based on Charles Sanders Peirce's work of the "logical basicality of implication or logical inclusion as asymmetrical and contingent" combined with a "weak" form of the Principle of Sufficient Reason for which one assumes that "all states of affairs must have sufficient reason."

Many formulations of naturalism, such as aspects of the "cosmic creation story" and even Big History (Christian 2013), are overly limited to a problematic materialism and deductivism, encouraging speculation beyond their proper bounds. All too often these materialist approaches convert a sound methodological principle into an implicit metaphysical claim, often represented as simply a "scientific" proposition (e.g., methodological reduction to reductionism; high levels of determination in physical relationships ("laws") to strong determinism; extending the application of "proof" in some limited areas of logic and mathematical physics to claims that all "rigorous" science is deductive). In contrast, Farzad Mahootian and I have shown how data-driven Big Data and new observational-inductive approaches are rapidly becoming very effective strategies for contemporary science and technology — strategies that break the "deductive" mold (Mahootian and Eastman 2009).

As a stark example of the appeal of "deductivism" in model development, Max Tegmark in his recent book *The Mathematical Universe* states that "our physical world not only is *described* by mathematics, but that it *is* mathematics, making us self-aware parts of a giant mathematical object" (2015, 6). George Ellis roundly criticizes Tegmark's neo-Platonism and his multiverse speculations — built on standard Big Bang cosmology and some mathematical wizardry — in which "all possible mathematical structures are realized in some unobservable physical reality" (Ellis 2015). As noted above, such hyper-realism can be readily avoided by simply distinguishing (ontological) possibility and actuality within the "real" as argued by Epperson and Zafiris (2013).

Although skeptical of any and all "-isms," I find Arran Gare's call for a new "speculative naturalism" to be helpful for tentatively meeting most of the criteria that I see as needed for a new integration; namely, that it is well-based in philosophy and the history of thought, inclusive, and incorporates the best of science and both epistemology and metaphysics. Unfortunately, Quine's "naturalistic turn" may have, as Gare comments, represented "a capitulation of philosophy to mainstream reductionist science" (2014). Similarly, Richard Rorty's "linguistic turn" (1967) may represent a capitulation to anti-realist anthropomorphism; however, there remains a worthy component of the analytic tradition (including the best of Quine's own works, with limited metaphysical commitments) that maintains a focus on sound methodology and philosophical analysis. Indeed, intersections of the analytic tradition with process thought have been reviewed extensively by George Lucas in his work *The Rehabilitation of Whitehead* (1990), and in the edited volume *Process and Analysis* (Shields, ed., 2003). Building on the best of the analytic tradition, it appears to me that a fully inclusive "speculative philosophy" should draw effectively on the following three complementary areas: analytic philosophy, the process tradition, and science. To these ends, one leading philosopher who embodies all three is Nicholas Rescher (University of Pittsburgh, emeritus) who has written extensively on all aspects of this complementary set of perspectives (e.g., Rescher 1994, 1996, 2007). Of course, there are many important elements of continental philosophy that would be beneficial as well, exemplified by the works of Gilles Deleuze (1988) and Jürgen Habermas (2008), among others. Another important source for complementary perspectives would be the works of leading thinkers in other cultural traditions such as Chinese or Indian. If he could have lived to witness these new developments, I suggest that my Professor Feigl would have likely worked to forward such an integration as well, yielding then a real "something more" philosophy.

## A POSSIBLE NEW FRAMING OF INTUITION

The importance of the process tradition is well illustrated by recent science-philosophy mappings and a possible new framing of intuition itself. Although Whitehead formulated his admittedly "speculative metaphysics" nearly a century ago, concurrent with the early stages of field theory, relativity, and quantum physics, that he was on the right track is strongly suggested by detailed mappings of its key elements to results arising from contemporary fundamental physics as carried out by Michael Epperson (2004) and Jesse Bettinger (2015). In spite of its errors and limitations, Whitehead's highly self-critical evolution of thought, combined with profound logical, mathematical, physical, and "metaphysical" intuition, enabled the creation of a philosophical/ metaphysical research program that is now bearing fruit around the world and in almost every field of research (see Weber and Desmond, eds., 2009).

Building on the key distinction, argued by Epperson and Zafiris (2013), of actuality and possibility within the real—and their associated distinction within applications of quantum physics (along with Bub (1999)) of Boolean and nonBoolean logic—I propose that the possibility "space" of the near future (with its nonBoolean logic) can be effectively tapped through quantum effects within certain complex biological systems. The neurobiological basis for such effects has been discussed in recent literature; e.g., Summhammer et al. (2012) and Bernroider (2012).

> Non-trivial quantum mechanics is clearly involved in some of the most important biological processes on the planet. As Schrödinger anticipated nearly 70 years ago, quantum mechanics looms large in life because the orderly structure of biological material extends from the macroscopic world right down to the atomic level. Here, biomolecules manipulate and exploit the quantum properties of fundamental particles such as electrons, excitons, protons and atoms: order from order. (McFadden 2013, 16)

With such a quantum basis, intuition may represent an enhanced ability to amplify the subtle information in the nonBoolean, near future, possibility space. This could as well be distinguished from the imagination, which may be referring to all possibility versus just the much more delimited compossible; i.e., that which is genuinely possible given all constraints and the given past. Indeed, at some future time, it may be possible to provide a partial measure of such intuition via some characterization of intuited realizations (e.g., musical passages created by Mozart) vs. some quantification of the full "possibility-space" compossible with given constraints (e.g., constraints of the musical genre, experience of the composer, etc.). Such quantification may be partly feasible through a focus on finite measures arising from a combination of quantum physical discreteness as well as through an application of finite mathematics (Van Bendegem 2012) and quantized measures for all appropriate modeling structures. In contrast, the standard approach has assumed the "causal closure of the physical" in combination with physical modeling that uses continuous functions and parameters subject to problematic infinities. If, instead, strict finitism was an adequate mathematics for the realm of actualizations (Boolean logic, quantization, discreteness), and in contrast the constructs of continuous functions, superpositions, etc., applied only to the near-future possibility space (nonBoolean, continuous), then problems of infinities can be resolved because they apply only to the possibility space, not to finite actualizations. In turn, physical models based on finitism can resolve many problems with contemporary physical models (Van Bendegem 2012; also this volume).

Given the various developments noted above, both scientific and philosophical,[8] I hypothesize that intuition, in part,[9] is associated with the amplification of subtle quantum effects that include both the actualized (Boolean, past-present domain) and the near future possibility space (nonBoolean, near future domain of compossibles). Such a framing of intuition would then provide both a partially-testable scientific basis for intuition, assuring consistency with existing physics, and yet at the same time would inevitably leave open a vast range of possibility that transcends any finite scientific analysis. This approach could eventually yield a basis of "reality" for intuition, subtle yet substantive, that leverages contemporary quantum experiments and interpretations, such as that of Epperson and Zafiris' Relational Realism (2013), that support the key distinction of the real in terms of both actualities and possibilities. Providing such a substantive basis for intuition could both exemplify a response to Arran Gare's call for a new mathematics (Gare, this volume), and potentially open a pathway for new foundations for meaning and value that better integrate the best of humanities and science.

## NOTES

1   See "supervenience" entry in the *Stanford Encyclopedia of Philosophy*.

2   See Joseph Needham (1956) for both this issue of linguistic comparison, and for substantial notes on the relevance of Whitehead's philosophy.

3   Quoted from paper by Pater Harris (2012) written in honor of philosopher James Bradley.

4   See Kraus (1998).

5   Logical causality denotes "causality that is logically conditioned" per compatibility conditions required by quantum physics (Epperson and Zafiris 2013, 58).

6   On the Free-Will Theorem, see http://www.informationphilosopher.com/freedom/free_will_theorem.html. As well, see Robert Valenza, "What does a particle know," this volume.

7   See Nobo (2004); also George Shields' detailed essay in which he concludes that "process panexperientialism seems to me to be a 'best explanation' relative to other available alternatives in the philosophy of mind." (Shields 2012). As well, a significant group of analytic philosophers of mind now appear to support this hypothesis (see Bruentrup 2016).

8   "The body in this process acts as a 'complex amplifier,' in which the experiences of the various parts of the body are enhanced en route to the central occasions of experience." (Griffin 2016, 259).

9   Complementary to the amplified quantum effects suggested here, a major component of 'intuition' is likely understandable in terms of classical signal processing as shown by Jesse Bettinger in his analysis of interoception and Whiteheadian perception (Bettinger 2016, this volume).

## REFERENCES

Bernroider, Gustav. "Neural Correlates of Higher Level Brain Functions." *NeuroQuantology* 10.2 (2012): 15–16.

Bettinger, Jesse. *The Founding of an Event-Ontology: Verlinde's Emergent Gravity and Whitehead's Actual Entities.* Claremont, CA: Ph.D. Thesis Claremont Graduate University, 2015.

Bruentrup, Godehard, ed. *Analytic Philosophy and Panpsychism.* Oxford: Oxford University Press, 2016.

Bub, Jeffrey. *Interpreting the Quantum World.* Cambridge: Cambridge University Press, 1999.

Christian, David. *Big History: Between Nothing and Everything.* New York: McGraw-Hill, 2013.

Clayton, Phillip. *The Predicament of Belief: Science, Philosophy, and Faith.* Oxford: Oxford University Press, 2011.

Cobb, John. *Theological Reminiscences.* Process Century Press. Anoka: MN, 2014.

Code, Murray. *Order & Organism.* New York: State University of New York Press, 1985.

—. *Myths of Reason: Vagueness, Rationality, and the Lure of Logic.* Trenton, NJ: Humanities Press, 1995.

Conway, John and Simon Kochen. "The Strong Free Will Theorem." *Notices of the AMS* 56.2 (2009): 226–32.

Deleuze, Gilles. *Bergsonism.* Brooklyn: Zone Books, 1988.

Deleuze, Gilles and Felix Guattari. *What is Philosophy?* Columbia University Press, 1996.

Desmet, Ronny. "Whitehead's Cambridge Training." In *Whitehead: The Algebra of Metaphysics*, eds. Michel Weber and Ronny Desmet, 91–125. Louvain-la-Neuve: Les Editions Chromatika, 2010.

Di Bernardo, Mirko. "Complexity and the Emergence of Meaning in the Natural Sciences and Philosophy." *Theology and Science* 13.2 (2015): 250.

Eastman, Timothy, and Hank Keeton, eds. *Physics and Whitehead: Quantum, Process and Experience.* Albany: State University of New York Press, 2004/2009.

Eastman, Timothy and Hank Keeton, eds. "Resource Guide for Physics and Whitehead." *Process Studies Supplements* Issue 6 (2004).

Ellis, George F. R. "Review of J. Leslie and R. Kuhn, eds. *The Mystery of Existence: Why is There Anything At All?*" *Theology and Science* 13.2 (2015): 261.

Epperson, Michael. *Quantum Mechanics and the Philosophy of Alfred North Whitehead*. New York City: Fordham University Press, 2004/2012.

Epperson, Michael and Elias Zafiris. *Foundations of Relational Realism*. Lanham: Lexington Books, 2013.

Fagg, Lawrence. *The Becoming of Time: Integrating Physical and Religious Time*. Durham: Duke University Press, 2002.

Ford, Lewis. *The Emergence of Whitehead's Metaphysics 1925–1929*. Albany: State University of New York Press, 1985.

Gare, Arran. "Speculative Naturalism." *Cosmos and History* 10.2 (2014): 300.

Griffin, David Ray. "Whiteheadian Physics: Its Implications for Time, Consciousness, and Freedom." In *Physics and Speculative Philosophy: Potentiality in Modern Science*, eds. T. Eastman, M. Epperson, and D. R. Griffin, 243-66. Berlin: De Gruyter/Ontos, 2016.

Habermas, Jürgen. *Between Naturalism and Religion: Philosophical Essays*. Cambridge: Polity Press, 2008.

Hare, Peter. "Whitehead." In *The Oxford Companion to Philosophy*, ed. Ted Honderich. Oxford: Oxford University Press, 1995.

Harris, Peter. "Creative Order: The Case for Speculative Metaphysics." *Analecta Hermeneutica* 4 (2012): 5.

Hartshorne, Charles. "Whitehead's Revolutionary Concept of Prehension." *International Philosophical Quarterly* XIX.3 Issue 75 (1979).

Kauffman, Stuart. "Res Potentia and Res Extensa Linked, Hence United, by Quantum Measurement." In *Physics and Speculative Philosophy: Potentiality in Modern Science*, eds. T. Eastman, M. Epperson, and D. R. Griffin, 47-51. Berlin: De Gruyter/Ontos, 2016.

Krauss, Elizabeth. *The Metaphysics of Experience: A Companion to Whitehead's Process and Reality*. New York City: Fordham University Press, 1998.

Lucas, George R. *The Rehabilitation of Whitehead: An Analytic and Historical Assessment of Process Philosophy*. Albany: State University of New York Press, 1990.

Mahootian, Farzad and Timothy Eastman. "Complementary Frameworks of Scientific Inquiry." *World Futures* 65 (2009): 61–75.

Mahootian, Farzad and Tara-Marie Linné. "Jung and Whitehead:

An Interplay of Psychological and Philosophical Perspectives on Rationality and Intuition." In *Rational Intuition: Philosophical Roots, Scientific Investigations*, eds. Lisa Osbeck and Barbara Held. Cambridge: Cambridge University Press, 2014.

McFadden, J. "Making the Quantum Leap." *Biologist* 60.2 (2013): 13–16.

McHenry, Leemon B. *The Event Universe: The Revisionary Metaphysics of Alfred North Whitehead*. Edinburgh: Edinburgh at the University Press, 2015.

Needham, Joseph. *Science and Civilization in China, Vol. 2 History of Scientific Thought*. Cambridge: Cambridge University Press, 1956.

Nobo, Jorge. "Whitehead and Quantum Experience." In *Physics and Whitehead*, eds. Timothy Eastman and Hank Keeton. Albany: State University of New York Press, 2004.

Puntel, Lorenz. *Structure and Being: A Theoretical Framework for a Systematic Philosophy*. University Park: Pennsylvania State University Press, 2008.

Raggett, Simon. *Consciousness, Biology and Fundamental Physics*. Bloomington: AuthorHouse, 2012. (Including a discussion of works by Gustav Bernroider: 93–95.)

Rescher, Nicholas. *A System of Pragmatic Idealism. Vol. III. Metaphilosophical Issues*. Princeton: Princeton University Press, 1994.

Rescher, Nicholas. *Process Metaphysics: An Introduction to Process Philosophy*. Albany: State University of New York Press, 1996.

Rescher, Nicholas. *Process Philosophical Deliberations*. Frankfurt: de Gruyter/Ontos-Verlag, 2007.

Rorty, Richard. *The Linguistic Turn*. Chicago: University of Chicago Press, 1967.

Rose, Philip. "Relational Creativity and the Symmetry of Freedom and Nature." *Cosmos and History* 1.1 (2005): 3–16.

Ross, Stephen D. *Perspective in Whitehead's Metaphysics*. Albany: State University of New York Press, 1983.

Shields, George, ed. *Process and Analysis: Whitehead, Hartshorne, and the Analytic Tradition*. Albany: State University of New York Press, 2003.

Shields, George. "Whitehead and Analytic Philosophy of Mind." *Process Studies* 41.2 (2012): 287–336.

Shields, George. "A Logical Analysis of Relational Realism." In *Physics and Speculative Philosophy: Potentiality in Modern Science*, eds.

T. Eastman, M. Epperson, and D. R. Griffin, 127–40. Berlin: De Gruyter/Ontos, 2016.

Stengers, Isabelle. *Thinking with Whitehead*, Cambridge, MA: Harvard University Press, 2011.

Summhammer, Johann, Vahid Salari, and Gustav Bernroider. "A Quantum-Mechanical Description of Ion Motion Within the Confining Potentials of Voltage-Gated Ion Channels." *Journal of Integrative Neuroscience* 11.2 (2012): 123–35.

Tegmark, Max. *The Mathematical Universe*. New York City: Vintage Books, 2015.

Unger, Roberto, and Lee Smolin. *The Singular Universe and the Reality of Time*. Cambridge: Cambridge University Press, 2014.

Van Bendegem, Jean Paul. "A Defense of Strict Finitism." *Constructivist Foundations* 7.2 (2012): 141–49.

Weber, Michel and Will Desmond. *Handbook of Whiteheadian Process Thought. Volumes 1 and 2*. Frankfurt: De Gruyter/Ontos-Verlag, 2009.

White, Alan. *Toward a Philosophical Theory of Everything: Contributions to the Structural-Systematic Philosophy*. London, NYC: Bloomsbury Publishing, 2014.

# 3. Whitehead on Intuition: Implications for Science and Civilization

FARZAD MAHOOTIAN

## AN ECOLOGICAL TURN

Whitehead re-envisioned science and philosophy in *Science and the Modern World,* culminating his initial efforts to revise physics (PNK, CN, R). Rather than attempting to justify Whitehead's philosophical approach, my aim here is to look for its potential application to a recent ecological turn in contemporary science and philosophy.

Whitehead proposed a more concrete worldview than modern science, one more consistent with human experience. Accordingly, he replaced the key abstraction at the core of the materialistic world view, namely "matter," with concrete self-determining, self-organizing entities, all constituted by and with a perspective on their given environments. Such "'Actual entities'—also termed 'actual occasions'—are the final real things, of which the world is made up. There is no going behind actual entities to find anything more real" (PR 18). These "final" units emerge from their ecological contexts through a process of selective exclusion, inclusion, combination, and modification of vectors of "feeling." The resultant intersection of these vectors is a uniquely self-organized "experience": an actual entity. Whitehead conceives each actual entity as an experiencing subject, where feeling is a metaphor for

their *selective responsivity* to one another, a responsivity through which
subjects are co-constituted and interconnected. Analogous to the way
that Sir Richard Attenborough characterizes life on Earth ("Each living
creature is an answer to the question 'What is Life?'") each actual entity
is—not has, but *is*—a unique standpoint on the universe as a whole.
Whitehead puts it this way: each entity embodies and expresses how the
world feels from its perspective, and is therefore a *valuation* of the whole
from the standpoint of a part. Indeed, he holds that the act of valuation
is "the very texture of realization in itself" (SMW 93).

Whitehead made feeling into a technical philosophical term that
retains the richness of its ambidextrous vernacular, connoting both
objective and subjective reference. Such an ambiguity enables reference
to world and self in the same breath, as when Whitehead says that
feelings are *vectors,* and when he suggests that the complexity of *value*
depends on the complexity of the experiencer. Valuation reaches both
inward and outward because each actual entity unifies the confluence
of environmental vectors and, in its turn, that entity infects its environment with valuations for the next generation of self-organizing entities.
On this basis, Whitehead can declare the central proposition of his
process ontology in one short sentence: "The many become one and are
increased by one" (PR 21).

## INTUITION

In Whitehead's process philosophy, as in the analytic psychology of Carl
Jung, *feeling* is an evaluative, selective, and thus *rational* function.[1] But
what constitutes rationality for Whitehead? We shall see (next section)
that he addresses this question directly with his "ontological principle,"
a key category of explanation within the "Categoreal Scheme" of his
speculative cosmology, *Process and Reality.*[2] According to this principle,
seeking a reason means seeking an actual entity, or a nexus of them. In
place of the term "feeling," Whitehead coined the term, "prehension" in
order to avoid promoting the impression that the actual constituents of
the world are cognitively apprehending their environments and making
conscious decisions about it. He distinguished two kinds of prehension:
"physical" and "conceptual" (PR 86). "Physical prehensions" are the

primary affective and unconscious apprehensions of the environment. "Conceptual prehensions" are constituted by processes of abstraction, contrast, and combination, i.e., the recognition of patterns and patterns of patterns, their replication, modification, and invention.[3] In a minority of entities who are complex enough to engender the "higher grades" of experience, consciousness emerges, intervenes, and enjoys its experience of the environment and even itself.

*Intuition,* for Whitehead, is a form of judgment on the threshold of consciousness. He divided what he called "intuitive judgments" into three classes: *affirmative, negative,* and *suspended.* The first two address prehensive conformation (affirmatively or negatively) with external data. The third form of intuitive judgment neither affirms nor denies conformity with the data. It *entertains* propositions (to use Whitehead's term) but *does not judge* them (PR 270). This *"suspense"* form of intuitive judgment most closely corresponds to C. G. Jung's idea of *active* intuition and C. S. Peirce's abductive inference in that it is creative and speculative, originary, and not reproductive. Whitehead believed that this active mode of intuition creates novelty and contributes to the creative advance, in Bergson's language, of the universe. Similarly, Peirce's abductive inference originates the premises of scientific inquiry, thus new potential knowledge.

Whitehead considered *consciousness* to be interstitial: it is "located" between the biological cells rather than "in" them. This interstitiality is structurally similar to active, i.e., nonjudgmental, intuition, which is suspended between a pair of binary opposites, i.e., *affirmative/negative* judgment. However, consciousness is suspended between *multiple* modes of order, i.e., in the interstices of the complex adaptive system of neurons and other substrates of consciousness.[4] The body's neural net offers a multiplicity of *potential* responses to an ever-changing flow of events, thus enlivening consciousness with requisite variety as it envisions and flows toward unborn futures. Whitehead's understanding of consciousness is an exemplary demonstration of how *possibility* is as relevant as *actuality* to shaping reality in his system of thought.

Whitehead's views on intuition and his critique of scientific thinking are captured in three place-based metaphors coined in *Science and the Modern World*. First, Whitehead criticizes the *provincialism* (vii) of science, implying that it is nearly blind to anything that lies beyond the boundaries of its established paradigms. Furthermore, it lacks a strong self-reflexive awareness of its own provincialism, and any views that deviate from its standards. Second, the fallacy of *misplaced concreteness* indicates that the concreteness of reality has been *temporarily lost*, that the grasp of scientific thinking on the concrete world is tenuous and removed. Third, the fallacy of *simple location* indicates a specific way of confusing abstractions for concrete realities, and vice versa. For example, the universe of classical physics assumed without question the infinite divisibility of, (1) space into points (that occupy no space), (2) time into instants (that take no time to elapse), and (3) matter into atomic point masses (located at a point of space, for an instant). Provincialism and the fallacies of simple location and misplaced concreteness express the conflation of what one posits and what exists, i.e., an ignorance of the status of abstractions, and of habits of intuition that have sunk below the threshold of consciousness.

## RATIONALITY

In the very opening chapter of *Process and Reality*, Whitehead says, "The methodology of rational interpretation is the product of the fitful vagueness of consciousness" (PR 15). And, in one of the last passages that he added to *Process and Reality* (Ford 1985), he again notes that the ultimate categories of his ontology, namely, the "ultimate notions of 'production of novelty' and 'concrete togetherness' are inexplicable either in terms of higher universals or in terms of the components [i.e, prehensions] participating in the concrescence . . . *The sole appeal is to intuition*" (PR 22, emphasis added). Whitehead here points beyond the inadequacy of language to the ultimate shortcomings of any system of thought, even one with a full range of concepts extending from the most concrete togetherness to the most abstract, most general sense of creativity. Thought and language can only lure, stimulate, and seduce one into taking an "imaginative leap." Indeed, Whitehead's entire system is one big imaginative leap crystallized in the medium of a philosophical language supersaturated with neologisms.

Concepts that are foundational to his discussion of rationality and intuition occur in the important first chapters of *Process and Reality* on speculative philosophy and the categoreal scheme of the philosophy of organism. As noted earlier, Whitehead's ontological principle has special relevance to his definition of rationality:

> That every condition to which the process of becoming conforms in any particular instance has its reason *either* in the character of some actual entity in the actual world of that concrescence, *or* in the character of the subject which is in process of concrescence. This category of explanation is termed the 'ontological principle.' . . . This ontological principle means that actual entities are the only *reasons;* so that to search for a *reason* is to search for one or more actual entities. (PR 24, 18th Category of Explanation, emphasis in the original)

By placing rationality on the basis of the self-transforming actual entity ("the subject which is in process of concrescence"), we enter a world radically different from that in which classical logic lives. Both the processual nature and the aesthetic character of Whitehead's cosmology are shaped by the position he takes with regard to logical foundations.

In his *Myths of Reason,* Murray Code displays the failed efforts of 19th and 20th century logicians to eliminate vagueness from logic and science. Contrary to this trend, C. S. Peirce and Whitehead address vagueness explicitly and challenged the centrality of the principle of identity, i.e., A = A. But what could be simpler than this principle? Whereas Frege, Quine, and armies of logicians employ this principle without questioning its solidity as foundation, Whitehead uses it as an Archemedian pivot to put logic and ontology in motion. Identity is not a given. In the actual world there is no stasis, only temporary, contingent stability: endurance. What endures is achieved, not given: it *becomes* then passes away. By rendering the key elements of his categoreal scheme dynamic, Whitehead creates an *ontological basis* for the affirmation-negation contrast that is so relevant to his notion of intuitive judgment.

His 21st and 22nd categories of explanation take on the principle of identity directly. According to the principle expressed in these categories, in order to be itself, A must include *non-*A:

(xxi) An entity is actual, when it has significance for itself. By this it is meant that an actual entity functions in respect to its own determination. Thus *an actual entity combines self-identity with self-diversity.* (PR 25, emphasis added)

(xxii) That an actual entity by functioning in respect to itself *plays diverse roles in self-formation without losing its self-identity.* It is *self-creative*; and in its process of creation transforms its diversity of roles into one coherent role. Thus 'becoming' is the transformation of incoherence into coherence, and in each particular instance ceases with this attainment. (PR 25)

In the 21st category of explanation, we see the tensional dynamic of self and not-self: identity emerges by the combination and integration of A and not-A. In Hamlet's world it was a case of to be or not to be; in Whitehead's it is to be and not to be. In the 22nd category, the dynamism is further specified in two ways. Whitehead's complementarity between A and not-A is driven by A but grows out of not-A: coherence grows out of incoherence. In the Daodejing, the complementarity of yin-yang is not symmetrical: both in word and spirit, this classic of Chinese philosophy favors yin over yang.[5] The asymmetry of Whitehead's system favors identity, but this identity is neither static nor given: it arises, as Heraclitus said, in the tension of opposites: the actualized givens vs. the not-yet (potential). The identity is not given in advance, it is created/discovered in process; the analogy with improvisational performance is operative, especially in the case of complex entities where attention to contrasts generates interesting products.

The actual entity is a process of self-formation; I prefer to call it *self-organization* because this enables me to draw on the metaphors of non-equilibrium dynamical systems. We may note that "the transformation of incoherence into coherence," in addition to being a reference to Heraclitean "order out of chaos," is descriptive of Prigogine and Stengers' (1984) eponymous book. In a series of groundbreaking articles, Joseph Earley[6] has applied the idiom of non-equilibrium chemical reaction systems to the task of thinking through Whitehead's cosmology and metaphysics and vice versa. In what follows, I take a similar approach and apply Whitehead's philosophy of intuition and rationality to understand the practice of scientific thinking.

## INTUITIONS AND PROPOSITIONS

To the extent that rationality can give reasons for things, it must abide by the ontological principle. Whitehead's answer to the question *"What is the role of intuition in rationality?"* has everything to do with how an actual entity *self-organizes* its satisfaction out of the raw physical "feelings" of its environment, i.e., the actual world out of which each actual entity creates itself. The process has several phases, depending on the complexity of the actual entity in question. The initial phase is the selection of environmental data for inclusion (as positive prehensions) and exclusions (negative prehensions). Further phases of integration, whose collective aim is to increase and maximize intensity and richness, propose comparisons and contrasts toward the final satisfaction of the emergent entity.

As it emerges from the cumulative wealth of vectors which constitute the past, the tensional dynamic between self-identity and self-diversity shapes the character of each actual entity's process of individuation. Its uniqueness arises from the selections it makes, its determination of potential contrasts between what is and what could be: it is the product of its decisions. Every one of its decisions, as an act of selection, is an act of valuation. Whitehead's claim, in *Science and the Modern World*, that *"value is the very texture of realization"* (93), is enshrined as the "Category of Conceptual Valuation" in *Process and Reality:* "From each physical feeling there is the derivation of a purely conceptual feeling whose datum is the eternal object determinant of the definiteness of the actual entity, or of the nexus, physically felt" (26).[7]

The actual entity's process of experience has four general phases: the initial physical phase, the conceptual phase, the propositional phase, and the final intellectual phase. In entities of appropriate complexity the fourth phase may give rise to intuitive judgment and conscious experience. What follows is an abbreviated sketch of the phases of experience that constitute an actual entity.

*Phase 1: Physical feelings* are the primary data of experience, initial vectors available to the nascent entity in its environment: the raw data of the past available to the present moment. *Phase 2: Conceptual feelings* are feelings of what Whitehead calls pure potentials — qualitative and mathematical — that characterize physical feelings. *Phase 3:*

*Propositional feelings* consist of the selection and arrangement of physical and conceptual feelings into propositional form. The key roles that propositions play in Whitehead's ontology are multifaceted, interesting, and complex. *Phase 4:* Two kinds of *intellectual feelings* arise from the integration of propositional feelings with physical feelings: conscious perception arises if the propositional component is perceptual, and intuitive judgment arises if its propositional component is imaginative.

For a glimpse into the complex interweaving of propositional feelings that form the fabric of a typical life, consider the essentially dynamic nature of the many social and historical contexts within which you live. Picture your place within a) the matrix of your family, b) your professional environment, c) the field of potentialities that form your career path. One lives in constantly shifting landscapes of interactions between the facts and the possibilities they afford. This is the propositional milieu of Whitehead's cosmos. At each moment, every entity is taking a stand, rendering a decision with respect to the multitude of relevant propositions, the vast majority of these decisions are—as our latest neuroscience affirms—not conscious. While propositional feelings may exhibit a bewildering array of possibilities, these may be divided into two very general types: i) a propositional feeling is *perceptual* if its conceptual component originates from the physical feeling with which it is integrated; ii) it is *imaginative* if its conceptual component originates from a physical feeling that is not part of the integration.

Whitehead notes that there are three kinds of intuitive judgment in response to a proposition, *affirmative, negative* and *suspended.* As we turn to suspended intuitive judgments, let us consider Whitehead's distinction between judgments and propositions:

> We shall say that a proposition can be true or false, and that a judgment can be correct, or incorrect, or suspended. With this distinction we see that there is a 'correspondence' theory of the truth and falsehood of propositions, and a 'coherence' theory of the correctness, incorrectness *and suspension* of judgments. (PR 191, emphasis added)

In this connection, let us consider perhaps the most oft-quoted statement that Whitehead makes about propositions: "in the real world it is more important that a proposition be interesting than that it be true." This statement occurs within a fittingly interesting passage:

> The fact that propositions were first considered in connection with logic, and the moralistic preference for true propositions, have obscured the role of propositions in the actual world. . . . *The result is that false propositions have fared badly, thrown into the dust-heap, neglected.* But in the real world it is more important that a proposition be interesting than that it be true. The importance of truth is, that it adds to interest. (PR 259, italics added)

Whitehead's view of propositions bears on analytic philosophy's "problem of counterfactuals." George Lucas holds that Whitehead's propositions fall "midway between the conceptualist approach of Saul Kripke . . . and the realist approach to factuals and counterfactuals [of] D. K. Lewis" (Lucas 1989,142). Whitehead leans more toward a realism that encompasses conceptuality by according it ontological status, without at the same time committing the fallacy of asserting that verbal statements are adequate expressions of a propositions (PR 11, 13). As "lures for feeling," Whitehead's propositions act as final causes not merely for syllogistic conclusions, but for actualizations of potential matters of fact. Whether the particular matters of fact are known or unknown, conceived or not, their immediate potential existence does not hinge on whether they have been expressed clearly, or even correctly in words. As lures for feeling, "they remain metaphors mutely appealing for an imaginative leap" regardless of how far the "elements of language be stabilized as technicalities" (PR 4).

Propositions already function the way metaphors do in the sense that they contrast *how things could be* with *how things are*. We may define explicitly *metaphorical* propositions as a subclass that edges intuition into consciousness. Metaphorical propositions, in addition to being lures for feeling, draw conscious attention to their metaphorical nature. That is, they redirect attention "backwards," as it were, *away* from strict verbal formulation toward polysemy. Such a movement is often erroneously considered to be a retreat from scientific advancement. Science students follow a long period of memorization, manipulation and mastery of technical terminology which is achieved through a suppression of polysemic potential of language and, by extension, to a suppression of self-reflexive thinking. I submit that this is an error of pedagogical judgment because it reverses the process of learning that progresses from engagement (what Whitehead calls the "romance" phase)[8] to precision. In actual practice, scientific thinking takes the "backwards" direction whenever it has arrived at some sort of impasse, whether large or small. As the greatest scientists have often noted, this is anything but a setback, for when scientists reach an impasse they regain what Socrates considered his own most valuable asset: learned ignorance. It is precisely at this point that technical terms and formulas, those marvels of precision, no longer function to advance the course of thinking, and instead their employment produce incorrect or even meaningless results. The impasse heralds the opportunity for a renewal of insight, a fresh path, a new angle of approach, a new cause for consideration of what might be.

Scientists typically deal with an impasse with an attitude similar to the initial romance phase of learning. In both cases there is entirely appropriate intense imaginative and aesthetic engagement with the issue at hand. For the scientist, there is a complementary, if temporary, lack of useful precision. Suitably unstuck from their insistent corner of the storehouse of specialized concepts, terms, and practices, in search of a new approach to the impasse, the inquirer may recur to musing (as Peirce called it) and metaphorizing. The journey away from one's cherished specialization inevitably leads back to it, but with refitted concepts and fresh eyes. The return is not by the same path: one returns to the impasse with new patterns of interaction, new forms of precision, testing and generalization. According to Whitehead, "suspended judgments are weapons essential to scientific progress" (PR 275), and "it is the task of the inferential process to convert a suspended judgment into a belief, or a disbelief" (PR 272).

## ON THE TRUE UTILITY OF FALSE PROPOSITIONS: A CONCRETE CASE

An especially compelling concrete example that justifies the philosophical defense of false propositions is to be found in the field of medical decision-making. Researchers at the Memorial Sloan-Kettering Cancer Center, specializing in the comparative study of cancer diagnostics and treatment techniques, have found that proper estimation of diagnostic test efficacy requires a comparative analysis of both true and false positive, and true and false negative test results. The practice of medicine deals with highly complex systems capable of a staggering variety of responses to pharmacological agents. This is one of the reasons why the efficacy of a given medication varies considerably in any given population. The same holds true for the reliability of diagnostic tests, as these regularly produce a significant number of false negatives (cases where the test doesn't register a disease agent when one is present) and false positives (the test registers a disease agent when one is not present). Partly because of the bewildering and dynamic variability of organismic chemistry, false positives and negatives cannot generally be eliminated from diagnostic testing, therefore they must be taken into account. Statistical methods for doing so have been in use since diagnostic tests were developed, and novel methods, such as "decision curve analysis," use cost-benefit analysis to give different mathematical weights to different types of error.[9] Accounting for false results—whether positive or negative—is crucial to analyzing the patient's risk.

Generalizing this principle, we may assert that careful philosophical, scientific, and technological accounting of false propositions is indispensable to understanding and interacting with a world of dynamically open systems. Whether produced by human or nonhuman agents, both true and false propositions must engage the interest and scrutiny of philosophy of science, technology, and engineering. Such ideas have been broached by philosophers before, but not (to my knowledge) in a manner that suggests the specific methodological prescriptions of much practical utility. For the most part, philosophical interest in false propositions has been directed at correcting, preventing, or eliminating them. This certainly sounds like a rational endeavor until one considers the ineradicable variability inherent in sampling populations for a given trait or property . . . especially if the population in question is as complex and as variable as living organisms. False propositions cannot be eliminated in fact, i.e., from the world; they can only be eliminated from serious consideration; and indeed this is a historically dominant bias inherent in Western philosophy.

Ian Hacking and other members of the Stanford school of philosophy of science analyzed the self-vindicating nature of scientific thinking and experimental design. Their intent was to lay bare the actual practice of science, which is so often at odds with the public image held by many laypeople and scientists. In particular, the dogmatic belief in absolute scientific objectivity and truth does great disservice to a realistic understanding of the ineradicable variability of scientific practice. Karl Popper and his school of philosophy of science gave falsification a fair shot, albeit only for the sake of eliminating false propositions. The effort to *control* conditions of observation guides the design of scientific instruments and experiment—this makes all science ultimately self-vindicating. But self-vindication is never completely successful: the impossibility (both in principle and in practice) of complete control over all pertinent experimental conditions guarantees openness, therefore surprise, new discoveries, and inventions.

While much research is occupied with tightening gaps and reducing an experimental system's openness, the alternative strategy, exemplified by statistical approaches to understanding diagnostic tests, is to work with the system's inevitable openness to error. The work that Peirce and Whitehead did over 100 years ago on the logic of scientific reasoning is more concretely applicable to clinical practice and applied science than the misplaced effort to achieve causal closure.[10]

When it comes to science and technology, we must consider that there are a great many scientists whose research involves many similar activities, like calibration. Nevertheless, characteristic practices of their specific areas of research vary greatly from field to field, and subfield to subfield. It is in these types of situations that affirmative and negative judgments, correctness and incorrectness, are vital. But there is more to science than this. Whitehead's three educational phases—romance, precision, and generalization—are relevant in science, too. There is a continuity of becoming between the romantic phase which is explicitly driven by intuition, the engagement phase in which intuition seeks the guidance of inference to guide its investigations, and the precision phase which largely purges intuition until an impasse, crisis, or other issue arises. The impasse throws the investigator "back to the drawing board,"in other words, back to the intuitive sources of inspiration and insight.

For Whitehead, intuition is a source of novelty and the creative advance of science and civilization. In *Adventures of Ideas*, Whitehead notes that human imagination is refreshed by its "recurrence to the utmost depths of intuition" (159), and indeed this is what most people consider to be a typical idea of what intuition is "good for." The context that Whitehead gives for this statement provides a deeper insight into the relationship between intuition and rationality within the natural life-cycle of ideas. While some see intuition as superseding or transcending "normal" rationality altogether, others see intuition as the highest form of thought that rationality strives to achieve. Although Whitehead's characterization of the relationship is somewhat complicated in its details, this passage captures the primary sense of it, as superseding rationality at one moment, only to be superseded by it at another:

> Systems, scientific and philosophic, come and go. Each system of limited understanding is at length exhausted. In its prime each system is a triumphant success; in its decline an obstructive nuisance. The transitions to new fruitfulness of understanding are achieved by recurrence to the utmost depths of intuition for the refreshment of imagination. (AI 159)

Science and philosophy are often associated with, and even upheld, as paradigms of rationality. Whitehead affirms this tendency, but also maintains that these rational systems become obstructive on a periodic basis — as their histories reveal. Intuition is the salve at such times, revitalizing and repositioning the imagination and aiming its transformative function at rational stagnation whenever it arises. Every intuition is true, but limited.

## BIOMIMETICS — SCIENCE REIMAGINED, ECOLOGICALLY

Like Wordsworth, Whitehead had a deep reverence for nature, so a sense of wildness is never far from his imagination. He knew all too well that science and technology are inherently risky: "major advances in civilization are processes which all but wreck the societies in which they occur" (AI 88). We can certainly claim the birth of the information age as one such advancement. Our task is "the creation of the future, so far as rational thought, and civilized modes of appreciation, can affect the issue" (MT 233). Let us extend the role of rational intuition in science now that "the most delicate, anxious consideration of the aesthetic qualities of the new material environment" (SMW 196) is requisite. Whitehead's critique encourages us to ask questions about the status and possibilities of science and to consider what science could become if intuition, in its various capacities, were understood and openly cultivated in education. What if science were sensitive to "the infinite variety of vivid values achieved by an organism in its proper environment" (SMW 199)? If standard experimental design sought interdependent systems instead of artificially isolated ones? In the closing sections of this chapter, I will briefly survey steps that have been taken in these directions.

Molecular biology, currently one of the most popular and well-funded branches of the life sciences, remains largely mechanistic. Nevertheless, within various other branches of life science, especially in ecology, systems biology and bioinformatics, we find countless examples of the interdependent character of biological, chemical, geological and physical processes. Whereas in physics, time is considered a mere indexical or metric quantity, in other fields it appears to have a substantive character. Chemistry is rife with examples of this: for many classes of molecules, including those with biological significance, the process of synthesis (i.e., the *history* of their formation) determines their structure and properties. The upshot of a Whiteheadian revision of the mechanistic view of nature and its replacement with an organic view is an improved understanding that fosters more diverse and more significant capacities for interaction within natural and created environments. Specifically, our understanding of systemic, ecological relationships, together with a shift in attitude toward green products and processes, has opened a space for working *with* nature, rather than attempting to assume control of it. This shift is captured in biomimetics, a new trend in engineering and design.

Biomimetics signals a slight return to the premodern mood of the Renaissance magician who wished to be Nature's cooperator, rather than its dominator. This shift of attention is more profound than any specific technology or technique. It is a recognition of the value inherent in each entity, and, furthermore, it is a recognition that individual value is derived from the accumulation and selective constellation of values inherent in the environment over time. We are coming to realize that every encounter with nature discloses the integrity of the environment, of the individuals immersed in it, and of the potential integrities that could arise from their interactions. Significant developments of this trend can be found in bio-inspired products, consulting firms, and, more recently, university courses and textbooks on biomimetics (e.g. Yoseph Bar-Cohen 2006 ).

In a chapter on education titled, "Requisites for Social Progress," Whitehead offers two scenarios that anticipate this form of realization: 1) The location of the Charing Cross railway bridge, selected for the sake of efficiency, displays a complete lack of environmental aesthetic, and 2) an ideal factory concretely re-imagined as an organism designed to accommodate human and natural flows, not just ergonomic and economic ones. Regarding the construction of the bridge, Whitehead noted that "the assumption of the bare valuelessness of mere matter led to a lack of reverence in the treatment of natural or artistic beauty. Just when . . . the most delicate anxious consideration of the aesthetic qualities of the new material environment was requisite . . . art was treated as a frivolity" (SMW 196). The bridge is a symbol of this loss of value-consciousness. It brings into focus that which was most needed:

> A striking example of this state of mind . . . is to be seen in London, where the marvelous beauty of the estuary of the Thames, as it curves its way through the city, is wantonly defaced by the Charing Cross railway bridge, constructed apart from any reference to aesthetic values.

> The two evils are: one, the ignoration of the true relation of each organism to its environment; and the other, the habit of ignoring the intrinsic worth of the environment which must be allowed its weight in any consideration of final ends. (SMW 196)

Against this negative example, Whitehead provides a positive one: we may re-imagine a factory as an organism by attending to interrelated values of organism and environment.

> A factory, with its machinery, its community of operatives, its social service to the general population, its dependence upon organizing and designing genius, its potentialities as a source of wealth to the holders of its stock is an organism exhibiting a variety of vivid values. What we want to train is the habit of apprehending such an organism in its completeness. (SMW 200)

This, for Whitehead, would be an example of art in its most general sense: "The habit of art is the habit of enjoying vivid values." But this is just what modern education lacks.

> What is wanting is an appreciation of the infinite variety of vivid values achieved by an organism in its proper environment. When you understand all about the sun and all about the atmosphere and all about the rotation of the earth, you may still miss the radiance of the sunset. We want concrete fact with a high light thrown on what is relevant to its preciousness.

> What I mean is art and aesthetic education. It is, however, art in such a general sense of the term that I hardly like to call it by that name. Art is a special example. What we want is to draw out habits of aesthetic appreciation. According to [my] metaphysical doctrine . . . to do so is to increase the depth of individuality. (SMW 199)

Whitehead here advocates a mode of aesthetic education that is not satisfied by merely requiring university courses on aesthetic theory and art history, but a curriculum that emphasizes the aesthetic dimensions of any and every topic.

## CONCLUSION: AESTHETIC ADJUSTMENT VIA SPECULATIVE SCIENCE FICTION

Since the 1970s, philosophers of the environment (Baird Callicott, Frederick Ferré, Donna Haraway, Kristin Shrader-Frechette, Brian Norton, among others) and feminist philosophers of science (Evelyn

Fox Keller, Sandra Harding, Lorraine Code, and others) have striven to reconcile the disjoint perspectives of science and technology development with ecological and ethical theory. While philosophical arguments for a revision of science have become progressively more sophisticated in both epistemological and political expression, no dramatic shift of technological and scientific practice is discernable, other than the aforementioned trends toward "green chemistry" and biomimetics. Both represent a fraction of the chemical industry and engineering fields respectively. Though they seem to be on a growth curve at present, these are still early days.

Speculative science fiction progresses much more quickly, as beautifully exemplified in several parts of Neal Stephenson's 1995 *Diamond Age: A Young Lady's Illustrated Primer*. Stephenson's biomimetic vision shines with a sustained clarity throughout. His book opens with the matter-of-fact description of a seaside desalinization plant that resembles a giant-scale version of a stand of cala lilies wherein form follows function elegantly, organically, and ecologically:

> Source Victoria's air intakes erupted from the summit of the Royal Ecological Conservatory like a spray of hundred-meter-long calla lilies. Below, the analogy was perfected by an inverted tree of rootlike plumbing that spread fractally through the diamondoid bedrock of New Chusan, terminating in the warm water of the South China Sea as numberless capillaries arranged in a belt around the smartcoral reef, several dozen meters beneath the surface.

Biomimetics exemplifies organic models of interdisciplinary design and research. Bacteria have been manipulated to produce chemicals of interest (or to consume them, in the case of smaller oil spills); insects have been fitted with microcameras, and biomimetic surveillance microbots have been modeled after them. Can this be done with entire factories like the one evoked by Stephenson? Does the increase in scale matter? Perhaps it is true that transforming the factory into an ecologically positive part of the landscape is easier than making small scale entities. So what about scales beyond manufacturing? Can policy, for example, be rendered organic and biomimetic enough to take advantage of the flows of matter and energy of their ambient environments? Assuming for a moment that it is more than mere propaganda, can China's stated policy of an ecological civilization, a "beautiful China,"[11] become viable? It is very likely that China, or other nations, will pursue such ambitions for purely economic reasons, that is, only for the sake of a single "vivid value," rather than cultivating the variety of values, as Whitehead urged. In the likely scenario where economic value leads, certain natural constraints will necessarily assert themselves even if they are *not recognized* as values, but merely as inconveniences. Stephenson has foreseen this as well, for even as he describes the cala lily-shaped factory as a marvel of eco-engineering in the passage quoted above, he goes on to note that,

> One big huge pipe gulping up seawater would have done roughly the same thing, just as the lilies could have been replaced by one howling maw, birds and litter whacking into a bloody grid somewhere before they could gum up the works. But it wouldn't have been ecological. The geotects of Imperial Tectonics would not have known an ecosystem if they'd been living in the middle of one. But they did know that ecosystems were especially tiresome when they got fubared, so they protected the environment with the same implacable, plodding, green-visored mentality that they applied to designing overpasses and culverts. Thus, water seeped into Source Victoria through microtubes, much the same way it seeped into a beach, and air wafted into it silently down the artfully skewed exponential horns of those thrusting calla lilies, each horn a point in parameter space not awfully far from some central ideal. They were strong enough to withstand typhoons but flexible enough to rustle in a breeze. Birds, wandering inside, sensed a gradient in the air, pulling them down into night, and simply chose to fly out. They didn't even get scared enough to shit.

Biomimesis is here presented as a plodding but effective way to escape the inevitable damage sustained by natural and man-made systems by rendering these systems less arbitrarily human and more explicitly natural. Stephenson's statement about geo-engineers carries a grain of cynicism not shared by all of his characters. For example, the artists, artisans, and engineers envisioned in the book embody an aesthetic imagination that is creative and not merely seeking to avoid ecologically messy consequences. Process philosophy, ecological vision, and the de-centered co-creativity exemplified in biomimesis will not prevent messiness, but the messes may indeed be more natural. However, even at its best, biomimesis is still only a mode of applying science and technology, and not all human problems are susceptible of scientific and technological solution.

Whitehead spoke of the habit of "enjoying vivid values" in reference to a well-rounded education aimed at increasing a student's "depth of individuality." While that goal is not achievable by science and technology alone, it is reasonable to cultivate the kind of science education that enables students to appreciate "the infinite variety of vivid values achieved by an organism in its proper environment . . . with a high light thrown on what is relevant to its preciousness" (SMW 199). In this passage, Whitehead was advocating a mode of education that would be especially applicable to natural science and engineering, but could apply just as significantly to the social sciences, economics and political science.

## NOTES

1    Though they diverge in their use of terms, both thinkers describe feeling and intuition in remarkably similar ways. See Mahootian and Linné (2014).

2    Whitehead's metaphysics is presented in a number of his writings but nowhere as systematically as in his Gifford Lectures, *Process and Reality: An Essay in Cosmology*. The bold subtitle notwithstanding, the book's very first sentence states that "This course of lectures is designed as an essay in Speculative Philosophy" (PR 4). While the latter label more accurately represents his philosophical effort in general, the subtitle expresses Whitehead's preference to think of his work as a speculative cosmology rather than a "provisional metaphysic."

3    Such processes are routinely carried out by computers without any ascription of conscious agency to them. It is widely acknowledged by Bruno Latour and others that agency, but not conscious agency, is appropriately applied to the "materiality" of scientific endeavors, most obviously in the case of sensors and other forms of laboratory instrumentation.

4    See Jesse Bettinger's chapter, this volume.

5    For example, see translator's introduction, Red Pine (2009).

6   See his forthcoming paper.

7   Whitehead's term "eternal object" designates ideal patterns of determi-
    nateness, e.g., the equilateral triangle, whose role as potential to act as
    a "determinant of the definiteness of the actual entity" is noted in this
    category.

8   See his *Aims of Education*.

9   A. J. Vickers and E. B. Elkin (2006). Another study refers explicitly to
    the inadequacy of statistical methods to correct "verification bias" in the
    absence of a sufficient number of false negative results: A. M. Cronin and
    A. J. Vickers (2008).

10  The author gratefully acknowledges the guidance of epidemiologist and
    biostatistician Andrew Vickers at the Memorial Sloan-Kettering Hospital
    with respect to decision curve analysis—the statistical technique used
    in the comparative assessment of diagnostic results. The broad epistemic
    assessment of these techniques, and any attendant errors of interpretation,
    are solely the responsibility of the author. Interestingly, Vickers informed
    me that after deriving formulas for decision curve analysis, it came to
    his attention that C. S. Peirce had proposed a similar approach (in an
    1884 communication to the journal *Science*), which offers a novel way of
    calculating the cost-benefit analysis of community evacuation in response
    to potential tornado damage.

11  A relatively new Chinese government policy for sustainable development
    refers to the establishment of practices consistent with "ecological civ-
    ilization" in the People's Republic of China: "In his report to the 18th
    National Congress of the Communist Party of China, former Chinese
    president Hu Jintao, emphasized the importance of ecological progress
    and for the first time, wrote about the building of a 'beautiful' China in
    the national overall development plan" (China Daily, 23/5/13, 1–2). Philos-
    opher, Arran Gare extends process thought into the politics of ecological
    civilization. A special blend of process philosophy and Marxism enjoys a
    degree of popularity in contemporary Chinese academia, which currently
    hosts about more than thirty centers of process philosophy. Indeed, Gare
    (2011) notes the origin of the term "ecological civilization" seems to be the
    contemporary Chinese government. Although the concept was not always
    in official favor, it has been a central part of every high level government
    policy report since about 2006.

## REFERENCES

Code, M. *Myths of Reason*. Trenton, NJ: Humanities Press, 1995.

Cohen, Y. *Biomimetics: Biologically Inspired Technologies*. Boca Raton,

FL: CRC/Taylor & Francis, 2006.

Cronin A. M., and A. J. Vickers. "Statistical Methods to Correct for Verification Bias in Diagnostic Studies Are Inadequate When There Are Few False Negatives: A Simulation Study." *BMC Medical Research Methodology* 8 (2008): 75.3.

Earley, J. E. E. "An Invitation to Chemical Process Philosophy." In *Epistemology of Chemistry: Roots, Methods and Concepts*, ed. J.-P. Llored. Forthcoming.

Ford, L. *The Emergence of Whitehead's Metaphysics, 1925–1929*. Albany: SUNY Press, 1985.

Gare, A. "Toward an Ecological Civilization: The Science, Ethics, and Politics of Eco-Poiesis." *Process Studies* 39.1 (2011): 5–38.

Lucas, G. *The Rehabilitation of Whitehead: An Analytic and Historical Assessment of Process Philosophy*. Albany: SUNY Press, 1989.

Mahootian, F. & Linné, T.-M. "Jung and Whitehead: An Interplay of Psychological and Philosophical Perspectives on Rationality and Intuition." In *Rational Intuition*, ed. B. Held & L. Osbeck. Cambridge University Press, 2014.

Prigogine, I., and I. Stengers. *Order out of Chaos: Man's New Dialogue with Nature*. NY: Bantam, 1984.

Red Pine, *Taoteching*. CO: Copper Canyon Press, 2009.

Stephenson, N. *Diamond Age: A Young Lady's Illustrated Primer*. NY: Bantam, 1995.

Vickers A. J., and E. B. Elkin "Decision Curve Analysis: A Novel Method for Evaluating Prediction Models." *Medical Decision Making* 26.6 (2006): 565–74.

# 4. *Whitehead's Notion of Intuitive Recognition*

RONNY DESMET

According to Whitehead, the human psyche is a temporal stream of occasions of experience in which each present occasion is a creative synthesis of feelings of antecedent somatic and psychic occasions of experience. Each feeling has an objective content (what is felt) and a subjective form (how it is felt), and each synthesis of the highest grade (each conscious occasion of experience) is a process of becoming that Whitehead analyzes in terms of four phases: the initial physical phase, the conceptual phase, the propositional phase, and the final intellectual phase. Whitehead does not conceive of our intuition as a separate faculty but as constituted by our intuitive judgments, which are defined as intellectual feelings arising in the final phase of the process of becoming and involving imaginative propositions. Moreover, according to Whitehead, the intuition operative in mathematics and physics is mainly constituted by the intuitive judgments that we can call intuitive recognitions.

## WHITEHEAD'S THEORY OF FEELINGS AND INTUITION

Physical feelings are the initial unconscious feelings of antecedent occasions of experience. Conceptual feelings result from the unconscious recognition of pure potentials—qualitative and mathematical—that

95

characterize physically felt occasions of experience. Propositional feelings (which have propositions as objective content) are the still unconscious feelings that result from the integration of conceptual feelings
(reduced to predicative feelings, which have predicative patterns as
objective content) with physical feelings (reduced to indicative feelings, which have logical subjects as objective content). A propositional
feeling is perceptual if its conceptual component originates from the
physical feeling with which it is integrated, and it is imaginative if its
conceptual component originates from a physical feeling that is not part
of the integration. Finally, intellectual feelings are the conscious feelings
that result from the integration of propositional feelings with physical
feelings (the physical feelings from which their indicative components
derive). An intellectual feeling is a conscious perception if its propositional component is perceptual, and it is an intuitive judgment if its
propositional component is imaginative.

A conscious perception results from the integration of a perceptual
feeling with the physical feeling of antecedent somatic and psychic occasions of experience from which its components originate. The conscious
awareness of the perceptual component of a conscious perception is
usually clear, and is called (perception in the pure mode of) presentational immediacy. The conscious awareness of the physical component
of a conscious perception is usually vague, and is called (perception in
the pure mode of) causal efficacy. A conscious perception is always an
integration of its perceptual and its physical component. Hence, the
conscious awareness that is part of its subjective form is never pure
presentational immediacy, nor pure causal efficacy. It is always an integration of these pure perceptual modes, which is called (perception in
the mixed mode of) symbolic reference.

An intuitive judgment, like a conscious perception, can be conceived as a case of symbolic reference from the presentational immediacy of its imaginative component to the causal efficacy of its physical
component. But another distinction is equally important. An intuitive
judgment is affirmative if its subjective form includes definite belief
in its imaginative proposition, and is negative if its subjective form
includes definite unbelief in its imaginative proposition. An affirmative

intuitive judgment is very analogous to a conscious perception, because what might be (the propositional) and what is (the physical) are (at least partially) identified. But: "The triumph of consciousness comes with the negative intuitive judgment. In this case there is a conscious feeling of what might be, and is not" (PR 273). But: "It is a great mistake to describe the subjective form of an intuitive judgment as necessarily including definite belief or disbelief in the proposition" (PR 272). There is a third case: an intuitive judgment can be a suspended judgment. A suspended judgment is the entertainment of the contrast between what may be (the propositional) and what is (the physical). It is "dominated by indifference to truth or falsehood," and can be called "conscious imagination" (PR 275). According to Whitehead, "suspended judgments are weapons essential to scientific progress" (PR 275), and "it is the task of the inferential process to convert a suspended judgment into a belief, or a disbelief" (PR 272).

It is remarkable how close Whitehead is to analytic philosophers at this point where his notion of subjective form of feeling meets their notion of propositional attitude. For example, in "Creativity and Imagination," Berys Gaut writes:

> A suggestion mooted by several philosophers, and one I think is basically correct, is that imagining that such and such is the case, imagining that *p*, is a matter of entertaining the proposition that *p*. Entertaining a proposition is a matter of having it in mind, where having it in mind is a matter of thinking of it in such a way that one is not committed to the proposition's truth, or indeed to its falsity. In contrast, the propositional attitude of believing that *p* involves thinking of the proposition that *p* in such a way as to be committed to the proposition's truth. (272)

The role of the subjective forms of the feelings at play in a creative synthesis of feelings cannot be overestimated. The process of becoming of a conscious occasion of experience is not primarily controlled by consciousness. It "is a progressive integration of feelings controlled by their subjective forms" (PR 232). "Consciousness," according to Whitehead, "presupposes experience, and not experience consciousness. It is a special element in the subjective form of some feelings" (PR 53). In other words:

"Consciousness is the crown of experience, only occasionally attained, not its necessary base" (PR 267).

## INTUITIVE RECOGNITION IN MATHEMATICS AND PHYSICS

Let me turn to mathematics and physics now. "The general science of mathematics," Whitehead writes, "is concerned with the investigation of patterns of connectedness, in abstraction from the particular relata and the particular modes of connection" (AI 153). In other words: "mathematics is the study of pattern in abstraction from the particulars which are patterned" (ESP III). Accordingly, what drives mathematics from the perceptual level of natural patterns of particulars to the imaginative levels of increasingly abstract patterns of relatedness is pattern recognition. Mathematical intuition is all about pattern recognition, and the intuitive judgments most characteristic of mathematicians are intuitive recognitions of pattern.

Moreover, according to Whitehead, recognition is not only pivotal with respect to mathematics, but also with respect to physics. On that account Whitehead claims: "Recognition is the source of all our natural knowledge" (PNK 56). For Whitehead, recognition in physics ranges from the perceptual recognition of sense-objects as well as spatial, temporal, and causal patterns of relatedness of events to the imaginative recognition of the uniform spatio-temporal pattern of relatedness of all events and the contingent causal patterns of relatedness of gravitational and electromagnetic events. As discussed in the Introduction to this anthology, Whitehead disagreed with Hume's restriction of perceptual recognitions to sense impressions, that is, with Hume's exclusion of perceptual recognitions of spatial, temporal, and causal relatedness. Also, as discussed by Gary Herstein in Chapter Eight herein, Whitehead disagreed with Einstein's conflation of space-time geometry with field physics, that is, of the essential relatedness of nature with its contingent relatedness. "The real point," Whitehead writes, "is that the essential relatedness of things can never be safely omitted. This is the doctrine of the thoroughgoing relativity which infects the universe and which makes the totality of things as it were a Receptacle uniting all that happens" (AI 153–54).

When utilizing the word "recognition" in his early writings, Whitehead does not always make clear whether he is referring to the qualitative and mathematical recognitions giving rise to conceptual feelings, which are unconscious recognitions, or to perceptual and intuitive recognitions, which are intellectual feelings and, hence, conscious recognitions. Nonetheless, the distinction has always been important to him. In 1920, in *Concept of Nature* he already wrote:

> To call recognition an awareness of sameness implies an intellectual act of comparison accompanied with judgment. I use recognition for the non-intellectual relation . . . which connects the mind with a factor of nature without passage. On the intellectual side of the mind's experience there are comparisons of things recognized and consequent judgments of sameness or diversity. . . . I am quite willing to believe that recognition, in my sense of the term, is merely an ideal limit, and that there is in fact no recognition without intellectual accompaniments of comparison and judgment. But recognition is that relation of the mind to nature which provides the material for the intellectual activity. (CN 143)

Conscious recognitions involve complex processes of becoming in which unconscious recognitions (called "physical recognitions" in *Process and Reality*) provide the conceptual material for the emergence of intellectual feelings, ranging from the conscious perception of sense data to the conscious imagination of the causal patterns of relatedness which we call the laws of physics. To clarify Whitehead's notion of conscious recognition, any example of mathematical pattern recognition is appropriate. Alternatively, I could give the example of Whitehead's perceptual recognition of the relations of extension and cogredience between events, lying at the basis of Whitehead's uniform space-time geometry and of his rejection of Einstein's variably curved space-time geometry. I could also give the example of his imaginative recognition of the analogy of gravitation with electromagnetism, lying at the basis of his alternative laws of gravitation and electromagnetism and of his agreement with Einstein's claim "that there is no logical path to these laws, only intuition" (226), and that "we can only grasp [the laws of physics] by speculative means" (266).

## GESTALT ILLUSTRATION

Instead, however, I illustrate Whitehead's notion of intuitive recognition with an example that is easy and yet also illustrates the important role of the subjective forms of the feelings at play. The example involves a well-known *Gestalt* experiment. This is no coincidence. Like Whitehead, the *Gestalt* psychologists in Berlin were inspired by the anti-associationism and radical empiricism in late 19th, early 20th century philosophy and psychology as well as by the development of the theory of electromagnetism in physics. Like Whitehead, they were holistic thinkers who held that the whole field of human experience is more than the sum (the mere association) of isolated sense data, and that it is analogous to the whole field of electromagnetism, which is more than the sum (the mere distribution) of simply located charges.

The experiment is from *Gestalt* psychologist Wolfgang Köhler, who presented it in his 1929 book, *Gestalt Psychology*, as the maluma-takete-experiment. But it also figures in the 2011 book by neuroscientist Vilayanur Ramachandran, *The Tell-Tale Brain*, where it is presented as the bouba-kiki-effect. Ramachandran writes:

Look at the two shapes in the figure:

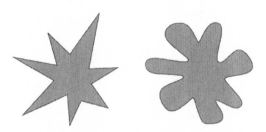

The one on the left resembles a jagged piece of shattered glass. The one on the right looks like a paint splat. Now let me ask you, if you had to guess, which of these is a "bouba" and which is a "kiki"? There is no right answer, but odds are you picked the splat as "bouba" and the glass as "kiki." I tried this in a large classroom recently, and 98 percent of the students made this choice. Now you might think this has something to do with the blob resembling the physical form of the letter B (for "bouba")

and the jagged thing resembling K (as in "kiki"). But if you try the experiment on non-English speaking people in India or China, where the writing systems are completely different, you find exactly the same thing. (108)

What would Whitehead say about the intuitive recognition of similarity between "bouba" and the rounded shape? In Whitehead's scheme each occasion of experience is an integration or concrescence (a growing together) of many feelings or prehensions, each of which has an objective content (what is felt) and a subjective form (how it is felt), and: "The process of the concrescence is a progressive integration of feelings controlled by their subjective form" (PR 232). With this in mind, we can take the intuitive recognition of similarity to involve an intuitive integration of the feeling "bouba" elicits and the feeling the rounded shape elicits, controlled by their subjective forms, and we can paraphrase a fragment from *Adventures of Ideas* to better understand this. Here we go:

> [The intuitive recognition of similarity arises because] the separate prehensions [the auditory and the visual feeling] of the two composite facts ["bouba" and the rounded shape] have been integrated so that the two objects stand in the unity of a contrast with each other. There is an intuition of a limited identity of pattern involved in the contrast of the diverse essences. In virtue of this identity there is a transference of subjective form from the feeling of one object ["bouba"] to that of the other [the rounded form]. What is appropriate to one is appropriate to the other. The intuitive recognition of 'that is so' is the subjective form including in itself justification of its own transference from the object on one side of the contrast to the object on the other side. (AI 242)

Of course, the same holds for "kiki" and the jagged shape. Also, in the global intuitive judgment that the rounded shape is a "bouba" and the jagged shape is a "kiki," intuitive recognitions of diversity are as important as intuitive recognitions of similarity. But the important point is that we are not yet conscious of *what* "bouba" and the rounded shape have in common at the moment of already becoming conscious *that* "bouba" and the rounded shape are similar. This example makes clear that our intuitive recognitions are not primarily the outcome of intellectual inference, but of emotional transference, that is, transference of subjective form.

## CONCLUSION

As a final point, I would like to add that, quite often, intuition leads mathematicians and physicists astray, and that it needs to be guided by logical thinking as well as calculations and measurements in order to avoid error. But it is a fallacy to conclude that logic or arithmetic or measurement can replace intuition, because all these guides require and reveal the intuition they have to guide. For example, mathematical logic requires and reveals the mathematical intuition it has to guide. It requires and reveals the intuitive recognition of pattern. In fact, all scientific activities are high grade conscious activities that presuppose intuitive judgments. To conclude, I paraphrase Whitehead:

> Science is rooted in intuitive judgments. That is the datum from which it starts, and to which it must recur. You may polish up intuition, you may contradict it in detail, you may surprise it. But ultimately your whole task is to satisfy it. (cf. OT 112)

## REFERENCES

Einstein, Albert. *Ideas and Opinions*. New York: Random House, 1954.

Gaut, Berys. "Creativity and Imagination." In *Imagination and Creativity*, ed. Michael Beaney, 268–93. UK: The Open University, 2005.

Köhler, Wolfgang. *Gestalt Psychology*. New York: Horace Liveright, 1929.

Ramachandran, V. S. *The Tell-Tale Brain: Unlocking the Mystery of Human Nature*. London: William Heinemann, 2011.

# 5. The Beauty of the Two-Color Sphere Theorem

RONNY DESMET

Gian-Carlo Rota wrote: "The beauty of a theorem is best observed when the theorem is presented as the crown jewel within the context of a theory" (130). The aim of this chapter is twofold. One: it offers an account of Whitehead's theories of process and beauty. This account aims at showing that his process-philosophical concept of beauty can be applied to mathematical theorems, and that his idea of the background dependency of beauty philosophically substantiates Rota's aphorism. Two: this chapter introduces the two-color sphere theorem; first as an isolated mathematical curiosity, and then in the context of one of the most famous theorems of quantum mechanics, the Kochen-Specker theorem.

This introduction in two steps aims at making the reader experience the truthfulness of Rota's aphorism. This chapter also has a very interesting spin-off: both "beauty," as in Whitehead's philosophy, and "spin," as in quantum mechanics, are instances of the notion of properties that do not exist prior to the aesthetic or measuring experience, but emerge from the actual processes of experience.

## WHITEHEAD'S PROCESS PHILOSOPHY

When browsing the philosophical writings of Whitehead, you will not find an essay entitled "Mathematics and Beauty." There is no such essay.

The two titles you will come across though are: "Mathematics as an Element in the History of Thought" (SMW 19–37), and "Mathematics and the Good" (ESP 97–113). At first sight, Whitehead did not explicitly relate mathematics to beauty, only to the history of thought and to the good. First appearances, however, can be deceiving. In both essays just mentioned, Whitehead also dealt with the aesthetics of mathematics (cf. SMW 24, 26-27; ESP 110-11). Moreover, and more fundamentally, Whitehead rejected the presupposition of a strict separation between thinking, desiring, and feeling, and between the corresponding ideals of truth, goodness, and beauty.

To understand Whitehead, we have to repudiate the faculty psychology that causes Kant's philosophy to divide into three critiques (cf. PR xiii). Central in Kant's three critiques are, respectively, cognition and science, conation and ethics, and affection and aesthetics. Like Whitehead, many of his British contemporaries rejected the Kantian split of the psyche into three separate faculties as well as the post-Kantian priority given to the faculty of knowledge and scientific understanding. But whereas most of them favored the faculty of desire to arrive at a unified psychology, Whitehead favored the faculty of feeling. Feelings are the starting point and the end point of Whitehead's unification. For him, feelings are not merely feelings of pleasure and displeasure; desires and thoughts are feelings too. Whitehead aspired to construct "a critique of pure feeling in the philosophical position in which Kant put his *Critique of Pure Reason*," and to supersede "the remaining *Critiques* required in the Kantian philosophy" (PR 113).

According to Whitehead, the human psyche is a temporal stream of occasions of experience in which each present occasion of experience—whether cognitive or conative or affective—is a creative synthesis of feelings of antecedent somatic and psychic occasions of experience. Each feeling has an objective content—*what* is felt—and a subjective form—*how* it is felt—and each synthesis of feelings is a process in which *how* many antecedent occasions are felt *controls* their integration into one new occasion of experience. "*How* an actual occasion *becomes*," Whitehead wrote, "constitutes *what* that actual entity *is*" (PR 23). In other words, the process of becoming of an occasion of experience is based on subjective forms—also called "emotional patterns" or "affective tones." The process of becoming, Whitehead claimed, "is a progressive integration of feelings controlled by their subjective form" (PR 232).

According to Whitehead, "the basis of experience is emotional" (AI 176). The basic glue of the psychic sticking together of many feelings into one new occasion of experience is made of unconscious affective tones, not of conscious cognitions, not even when the integration does give rise to a conscious and cognitive occasion of experience. "The principle that I am adopting," Whitehead wrote, "is that consciousness presupposes experience, and not experience consciousness" (PR 53). And he added:

> Consciousness flickers; and even at its brightest, there is a small focal region of clear illumination, and a large penumbral region of experience which tells of intense experience in dim apprehension. The simplicity of clear consciousness is no measure of the complexity of complete experience. Also this character of our experience suggests that consciousness is the crown of experience, only occasionally attained, not its necessary base. (PR 267)

The precognitive and even preconscious emotional pattern of feelings is the necessary base of their progressive integration into a new occasion of experience, even if ultimately this occasion is conscious or even cognitive.

Speaking of a process of progressive integration of feelings of past occasions suggests that this process is temporal and merely a matter of efficient causation. This suggestion is wrong. Even though Whitehead conceived of the psyche as a temporal macro-process of efficient transitions from past to present to future occasions of experience, he also held that the becoming of each individual occasion of experience is a non-temporal and purposive micro-process of creative synthesis of the felt past. In Whitehead's process philosophy, "there are two species of process, macroscopic process and microscopic process. . . . The former process is efficient; the latter is teleological" (PR 214). Also: the former process is temporal; the latter is nontemporal. The psyche—and imaginatively generalizing, even reality as a whole—emerges from a mixture of temporal and nontemporal processes, and of efficient and final causes.

The conception of the microscopic process of becoming as a progressive integration of feelings is merely the intellectual outcome of an ad hoc analysis. It results from a reflection on what is immediately given. According to Whitehead in *Process and Reality*:

> The conclusion is that in every act of becoming there is the becoming of something with temporal extension; but that the act itself is not extensive, in the sense that it is divisible into earlier and later acts of becoming which correspond to the extensive divisibility of what has become. (PR 69)

The authority of William James can be quoted in support of this conclusion. He writes: "Either your experience is of no content, of no change, or it is of a perceptible amount of content or change. Your acquaintance with reality grows literally by buds or drops of perception. Intellectually and on reflection you can divide these into components, but as immediately given, they come totally or not at all." (PR 68)

In an intellectual analysis of the process of becoming of a conscious occasion of experience, we can distinguish various stages of integration. In fact, Whitehead distinguished the physical, conceptual, propositional, and intellectual stages of integration. But in reality they cannot be separated. Likewise, in an intellectual analysis of feelings, we can distinguish the subjective form of feeling from its objective content. But in reality they are inseparable. As Whitehead wrote: "The intellectual disjunction is not a real separation" (PR 233). Each feeling is a unity of form and content, and each occasion of experience is a unity of feelings — simultaneously and inseparably.

## WHITEHEAD'S AESTHETICS

For Whitehead, the emergence of beauty is related to how a new occasion of experience becomes. Primarily, beauty emerges as an element of the integral subjective form of the feelings constituting a new occasion of experience. Its emergence depends on the perfection of the integral emotional pattern of the new occasion of experience. For the finite occasions of experience constituting our psyche, absolute perfection is not possible, but relative perfection is: "There are in fact higher and lower perfections" (AI 257). To clarify how higher perfection can be attained or inhibited, Whitehead distinguished two meanings of inhibition. I introduce these by means of two questions.

One: are as many feelings as possible included in the integration of feelings of antecedent occasions that gives rise to the new occasion? When this is the case, the variety of detail of the integral emotional pattern is optimal, and the highest possible type of perfection can be attained. When it is not, and feelings of antecedent occasions are excluded, the emotional pattern suffers from simplification, and only lower types of perfection can be attained. Whitehead used the term "anesthesia" for the inhibition of variety of detail of the integral emotional pattern by exclusion and simplification (cf. AI 256). Anesthesia does not necessarily derogate from perfection, but from going up in the hierarchy of types of perfection. There is another kind of inhibition, however, the kind that does derogate from perfection.

Two: are competing feelings of antecedent occasions integrated into effective contrasts? In other words: is the creative synthesis of a multiplicity of feelings of past occasions into the unity of a new occasion also transforming "diversities in opposition" into "diversities in contrast" (PR 348)? When this is the case, the integral intensity of contrast implies the promotion of the individual intensities of the diverse feelings. When it is not, and the diverse feelings give rise to discordant feelings, the integral intensity of discordance implies the mutual destruction of individual intensities. Whitehead used the term "aesthetic destruction" for the inhibition of effective contrasts in the integral emotional pattern by discordance and mutual destruction (cf. AI 256). Aesthetic destruction derogates from perfection. In fact: "The more intense the discordant feeling, the further the retreat from perfection" (AI 256).

As the emergence of beauty in the process of becoming of an occasion of experience is implied by the attainment of perfection of its subjective form, and as the optimal level of perfection is promoted by the decrease of anesthesia and aesthetic destruction, we can conclude that beauty is promoted by the increase of variety of detail with effective contrast. Whitehead also formulated this conclusion in terms of "massiveness" and "intensity." He wrote:

> The perfection of Beauty is defined as being the perfection of Harmony; and the perfection of Harmony is defined in terms of the perfection of Subjective Form in detail and in final synthesis. Also, the perfection of Subjective Form is defined in terms of 'Strength.' In the sense here meant, Strength has two factors, namely, variety of detail with effective contrast, which is Massiveness, and Intensity Proper which is comparative magnitude without reference to qualitative variety. But the maximum of intensity proper is finally dependent upon massiveness. (AI 252–53).

The last sentence of this quote makes clear that Whitehead's distinction between massiveness and intensity is again an example of an intellectual distinction, not of a real separation. In fact, it coincides with his distinction of two factors of pattern in the emotional pattern that is the subjective form — "its qualitative pattern and its pattern of intensive quantity" — of which he wrote: "these two factors of pattern cannot wholly be considered in abstraction from each other" because "the two patterns are not really separable" (PR 233). Instead of saying that beauty is promoted by the increase of massiveness (variety of detail with effective contrast), we might say that it is promoted by the heightening of intensity because such heightening arises when the diversity of antecedent occasions can enter explicit feeling as effective contrasts, instead of being excluded from explicit feeling, or instead of entering explicit feeling as discordant feelings (cf. PR 83). But it is safest to say that beauty is promoted by the increase of both massiveness and intensity, because discordant feelings can also be very intense, even though they imply the mutual destruction of the individual intensity of their component feelings. A peak of intensity can mean a peak of discord as well as a peak of harmony, "the height of Evil" as well as "the height of Beauty" (AI 261).

Whitehead agreed with the classical formulae of beauty: beauty arises from unity in diversity, and from simplicity in complexity. But the above account makes clear that beauty does not emerge in the process of becoming of an occasion of experience when unity is attained at the expense of diversity, and simplicity at the expense of complexity. With each new occasion of experience, with each new creative synthesis of feelings of antecedent occasions, unity emerges from diversity. Beauty, however, does not always qualify the integral emotional pattern of the new unification of the diverse feelings of the past. It only does so to the extent that in the process of unification both aesthetic destruction and anesthesia are overcome, and both intensity and massiveness are increased. As Brian Henning puts it:

> The zero of beauty, as Charles Hartshorne noted, is the zero of actuality. In determining itself, every occasion necessarily achieves some degree of beauty and is, to that degree, justified in its existence. Yet, it is still very much the case that an occasion of experience can fall short of the maximally unified diversity and balanced complexity open to it. It is, in this sense, ugly. (203)

The primary meaning of beauty has been given. It is the beauty of subjective form/emotional pattern/affective tone. Whitehead, however, also gave a secondary meaning of beauty. He did not only account for subjective or *how*-beauty, but also for objective or *what*-beauty:

> In the analysis of an occasion, some parts of its objective content may be termed Beautiful by reason of their conformal contribution to the perfection of the subjective form of the complete occasion. This secondary sense of the term Beauty is more accurately to be considered as a definition of the term 'Beautiful.' . . . 'Beautiful' means the inherent capability for the production of Beauty when functioning as a datum in a percipient occasion. When 'Beauty' is ascribed to any component in a datum, it is in this secondary sense. (AI 255–56)

So when beauty, for example, is ascribed to a work of art it is in the secondary sense that the work of art has the inherent capability, once it is felt by an art-lover, to heighten the perfection and produce the beauty of the integral subjective form of the corresponding occasions of experience (the occasions of experience that constitute the psyche of the art-lover *and* include the feeling of the work of art). The same holds for a mathematical theorem. It is beautiful if it possesses the inherent power to promote perfection and beauty of emotional pattern whenever felt by a mathematician. Generally speaking, an object is beautiful to whatever extent the subjective form of the feeling of it is beautiful.

## MOZART'S JUPITER SYMPHONY

Natural beauty can be evoked to illustrate the adequacy of Whitehead's aesthetics. Indeed, as Regine Kather writes:

> Nature has an aesthetical dimension. The unity of every organism is based on the integration of contrasting elements. Therefore unity does not mean uniformity. On the one hand, inconsistencies would destroy the unity of an organism and cause stress and suffering; on the other hand, the lack of contrasts is a sign of the lack of complexity and intensity of life. (194)

However, I stick to artistic beauty to highlight the applicability of Whitehead's theory of beauty because the transition from art to mathematics is smooth. In fact, many authors speak of the art of mathematics (cf. King). In "Beauty can save us: Mozart and Whitehead," Patricia Adams Farmer writes that "real beauty is not an abstract concept," but "an experience . . . that defies being tied down by language," and she adds:

> We might simply point to Mozart's Symphony No.41 in C—the "Jupiter"—and be done with it. After listening to the last movement of the Jupiter, having experienced the entrancing counterpoint melodies colliding into one another until you think your brain might explode with joy—well, what more can one say? (103)

But Farmer still resorts to words in an attempt to understand why Mozart's music is experienced as beautiful:

> In the last movement of the Jupiter, five separate melodies—not two or three, but five distinct melodies!—not only exist together but play together, tumbling toward each other and away again, daring one another, touching one another, challenging one another until, finally, they give up the chase and

simply sound together all at once. At once! The many become one in the final fugal coda. (107–08)

But this unprecedented fugue is no chaotic free-for-all. Harmony exists. But how? The secret is this: there is something at the center of the co-mingling melodies, a four-note motif or theme that somehow makes the whole thing work. This simple thematic melody allows the four other melodies to express themselves contrapuntally with a sense of meaning and connection, each melody playing off the motif melody with its own spirited voice. Without this motif (which pervades the whole symphony), all would be chaotic dissonance. (108)

Reduce the Jupiter to just whistling the theme, and this rude simplification will drastically lower the level of perfection. Arbitrarily mingle the competing melodies, instead of contrapuntally, and discordant feelings will most likely inhibit any experience of perfection.

"Beauty," Farmer writes, "can save us from our smallness and lack of vision and our impossible and impoverishing demands that life be limited to one key—and to hell with competing melodies" (106). And then she adds: "Beauty of this kind—Mozart's kind, the complex and contrapuntal kind—shines out from the center of the philosophy of Alfred North Whitehead" (106). Indeed, for Whitehead too, the perfection of beauty saves us from smallness and lack of vision, that is, from anesthesia and aesthetic destruction. Or, putting it the other way around, for Whitehead as for Mozart, complexity and counterpoint, that is, variety of detail and effective contrast, are avenues to heighten the perfection of beauty.

## EULER'S FORMULA

In 1988, David Wells published a list of 24 well-known mathematical theorems in *The Mathematical Intelligencer*, asking his mathematician readers to give each theorem a score between 0 and 10, expressing its beauty. In 1990 the results of this survey were published, and the theorem with the highest average score was the following formula, attributed to Leonard Euler:

$$e^{i\pi} + 1 = 0$$

In 2014, *Frontiers in Human Neuroscience* published the conclusions that Semir Zeki and his coauthors drew from their fMRI brain scans of 15 mathematicians who were rating mathematical formulae as beautiful, indifferent, or ugly. They showed not only that the experience of mathematical beauty involves the same part of the emotional brain as the experience of beauty derived from other sources, but also that the formula most consistently rated as beautiful was again the one that came out as the most beautiful in Wells' survey—Euler's formula.

A Whiteheadian account of why Euler's formula produces an intense experience of beauty for many mathematicians follows from comparing it with the four-note thematic melody of Mozart's Jupiter symphony. Indeed, hard as it is to experience for non-mathematicians, Euler's formula is like a simple mathematical melody that connects four others: the melody of the natural number system generated by the number 1, the melody of the real number system represented by the irrational number $\pi$; the melody of the complex number system produced by the imaginary number $i$; and the melody of the exponential and logarithmic functions based on the transcendental number $e$.

Each of the four numbers in Euler's formula—1, $\pi$, $i$, and $e$—is fundamental in a particular domain of mathematics, and yet they are brought together in a simple formula: take $e$ to the power of $i$ multiplied by $\pi$, then add 1, and the result is 0. Euler's easy formula unites a diversity of intricate number systems. "Euler's formula," François Le Lionnais wrote, "establishes what appeared in its time to be a fantastic connection between the most important numbers in mathematics" (128). And yet, its unification of this variety of numerical detail does not lead to any discordance of rational versus irrational, or real versus imaginary, or linear versus exponential, but turns these oppositions into effective contrasts. As Edward Rothstein put it: "It is as if some glimpse has been granted to us of another world which was previously hidden" (149).

The reason why Euler's formula does not produce any experience of beauty in most non-mathematicians is obvious. They are like non-musicians looking at the sheet music of the motif of Mozart's Jupiter symphony. They only see meaningless symbols, without hearing the thematic melody, let alone the other four melodies it connects. "People outside mathematics," Jerry King wrote, "may fail to see what the fuss is all about" (138). "Appreciation of mathematical beauty," Gian-Carlo Rota explained, "requires familiarity with a mathematical theory, which is arrived at at the cost of time, effort, exercise, and *Sitzfleisch* . . . Familiarity with a huge amount of background is the condition for understanding mathematics" (128–30). In fact, even mathematicians do not all share an identical background and, hence, cannot all equally appreciate Euler's formula.

## BEAUTY IS BACKGROUND DEPENDENT

Consequently, I was incorrect to claim on behalf of Whitehead that a beautiful work of art and a beautiful mathematical theorem have the inherent capacity to produce beauty of emotional pattern whenever felt by the art-lover or mathematician. This claim is in need of further qualification. Clearly, given a particular work of art, it is not true that every art-lover will experience it as beautiful at all times. A painting experienced in the 18th century as beautiful might not be experienced as such today; music experienced in the East as beautiful might not be experienced as such in the West; a poem experienced as beautiful by one literature graduate might not be experienced as such by another, who graduated the same year at the same university. Likewise, given a particular mathematical theorem, it is not true that every mathematician or every society of mathematicians will experience it as beautiful at all times because, say, the majority of the early 20th century mathematicians at Göttingen did.

Does this imply that beauty is purely subjective, that beauty is only in the eye of the beholder? The Whiteheadian answer reads: "No, beauty is objective as well; the individual, cultural, and historical differences of appreciation of beauty are dependent not merely on subjective evaluation, but also on differences in objective background." I will now unpack this concise answer.

Whitehead confirmed, again and again, the dual meaning of beauty:

There are two sides to aesthetic experience. In the first place, it involves a subjective sense of individuality. It is *my* enjoyment. . . . In the second place, there is the aesthetic object which is identified in experience as the source of the subjective feeling. (ESP 130)

But Whitehead also admitted that the thought "that there is a definite object correlated to a definite subjective reaction" is an abstraction. (ESP 130). He even spoke of "a violent abstraction," for every occasion of experience "has an infinitude of relations in the historic world and in the realm of form," while this abstraction only involves "a minute selection of these relations" (MT 89). Now and again Whitehead spoke of "pulses of emotion" (PR 163) instead of occasions of experience, and he wrote:

> The data for any one pulsation . . . consist of the full content of the antecedent universe as it exists in relevance to that pulsation. They are this universe conceived in its multiplicity of details. These multiplicities are antecedent pulsations, and also there are the variety of forms harbored in the nature of things, either as realized form or as potentialities for realization. Thus the data consist in what has been, what might have been, and what may be. (MT 89).

In other words, when we conceive of our present subjective experience of beauty as objectively caused by the work of art of which we are now conscious, and *only* by this work of art, we forget that *how* we feel the work of art at this moment in time is determined not only by this one abstracted objective cause, but by our subjective evaluation of the whole causal past of this occasion of experience. This does not imply that Whitehead wanted to ban all abstractions. Our finite intellectual feelings necessitate abstraction—"You cannot think without abstractions" (SMW 59; cf. also MT 123–24). It does imply, however, that Whitehead wanted to remedy our forgetfulness, and relativize our finite thoughts by putting them in the context of the infinite background from which they emerge. Hence he wrote:

> The finite intellect deals with the myth of finite facts. There can be no objection to this procedure, provided that we remember what we are doing. (MT 10)

> In our experience there is always the dim background from which we derive . . . (ESP 123)

That vast background of feeling is hardly touched by consciousness ... (MT 116)

It defies analysis by reason of its infinitude. (ESP 95)

We experience more than we can analyze. For we experience the universe, and we analyze in our consciousness a minute selection of its details. (MT 89)

This is the habitual state of human experience: a vast undiscriminated, or dimly discriminated background, ... and a clear foreground. (AI 260)

In sum, the differences of aesthetic experience, from moment to moment, from individual to individual, from community to community, and from era to era, are not purely subjective. They are caused by differences of objective background. The many cases in which one person's aesthetic judgment with respect to a mathematical theorem differs from another's do not add up to the claim that mathematical beauty is purely subjective, and in no sense objective. Rather, they imply that the emotional pattern of the conscious experience of a mathematical theorem is inseparable from its unconsciously felt background. In other words, the beauty of a mathematical theorem is not reducible to the mathematician's subjective evaluation, nor to the mathematical theorem's objective properties; it is inescapably background dependent.

Let me reiterate the latter claim with another Rota quote:

The beauty of a piece of mathematics is dependent upon schools and periods. A theorem that is in one context thought to be beautiful may in a different context appear trivial. . . . Many occurrences of mathematical beauty fade or fall into triviality as mathematics progresses. However, given the historical period and the context, one finds substantial agreement among mathematicians as to which mathematics is to be regarded as beautiful. . . . In other words, the beauty of a piece of mathematics does not consist merely of the subjective feelings experienced by an observer. The beauty of a theorem is an objective property on a par with its truth. . . . Mathematical beauty and mathematical truth share the fundamental property of objectivity, that of being inescapably context dependent. (126)

## THE TWO-COLOR SPHERE THEOREM

According to both Whitehead and Rota, experiencing mathematical beauty is background dependent. This can be demonstrated, for example, by mathematical theorems for which the minimal theoretical background sufficient to simply understand them is insufficient to also experience their beauty; theorems, that is, of which the beauty can only be felt when they are presented in a broader theoretical context. The remainder of this chapter focuses on one such theorem. In this section, I introduce the two-color sphere theorem with a minimum of resources. In the next one, I will indicate the relevance of the two-color sphere theorem in the context of quantum mechanics. I hope that the shift in my account from minimal mathematical background to extended scientific background will produce a transformation of the reader's experience of the theorem from a feeling of indifference to a feeling of beauty. Of course, I cannot control the reader's unconscious background, but I hope that he or she will feel an increase of aesthetic pleasure, and that this experience will lead him or her to affirm the truthfulness of the following aphorism of Rota: "The beauty of a theorem is best observed when the theorem is presented as the crown jewel within the context of a theory" (130).

The two-color sphere theorem is about coloring a sphere using only two colors, say, green and red. In general, when one is free to color as one pleases, this presents no difficulty. For example, I can paint one hemisphere of a football in green, and the other in red. However, by subjecting the coloring to a particular condition, and hence restricting my freedom of painting, it becomes not merely difficult, but even impossible. Indeed, the two-color sphere theorem holds that it is impossible to paint a given sphere s in green and red such that in the end the following holds: for each triplet of orthogonal symmetry-axes of s (that is, for each set of three straight lines through the center of s, which are mutually perpendicular), there is one of these three axes for which the two intersection points with the surface of s are green, and two for which the intersection points with the surface of s are red.

Observe that in order to satisfy this overall coloring condition, an infinite number of partial conditions must be satisfied, for there are an infinite number of different triplets of orthogonal symmetry-axes of s.

However, a stronger theorem holds, which implies the two-color sphere theorem and only involves a finite number of partial conditions. Let A be the set of 33 symmetry-axes of S determined by the 33 points drawn in the first figure below on the cube enclosing S, then this stronger theorem states that it is impossible to paint S in green and red such that for each triplet of orthogonal A-axes, there is one axis of which the intersection points with the surface of S are green, and two axes of which they are red.

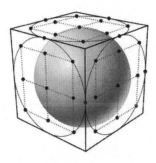

Because this second coloring condition only involves a finite number of partial conditions, a relatively simple case-by-case proof can be given of the second impossibility theorem. As a first case, I arbitrarily choose three orthogonal A-axes: the intersection points of the surface of S with one of the three axes, say X, must be green, and those with the other two axes, say Y and Z, must be red, as indicated in the next figure in shades of grey, where light grey stands for green and dark grey for red.

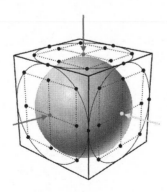

Then, as subsequent cases, I successively choose the other triplets of orthogonal A-axes of which X is also part. The coloring must be as indicated in the next figure (again: light grey = green and dark grey = red).

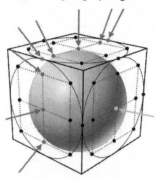

And so on. I continue to choose appropriate triplets of A-axes, and demand that the corresponding, partial coloring conditions hold, until I am stuck as shown in the next figure (again: light grey = green and dark grey = red).

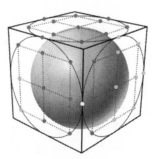

The intersection points of the surface of S with the symmetry-axis of S determined by the white point must be painted in green as well as in red in order to satisfy all partial conditions. But this is impossible, which proves the stronger impossibility theorem and, consequently, the weaker or two-color sphere theorem as well.

Notice that the sequence of handling cases and axes in the above case-by-case proof is quite irrelevant because of the symmetry of the setup. This symmetry is made more obvious by means of the next figure, in which the 33 axes of A are identified as the symmetry axes of the dark grey cube, the medium grey cube, and the light grey cube, which are obtained by rotating the original white cube through 45° about the X-, Y-, and Z-axis.

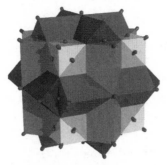

Chances are the reader, like me, did not experience any beauty when studying the forgoing account of the two-color sphere theorem and its proof (except—perhaps—when looking at the last figure above). But let me now try to increase our aesthetic appreciation of the two-color theorem by taking into account its scientific background, namely, by putting it to work in the context of the proof of the famous Kochen-Specker theorem.

## THE KOCHEN-SPECKER THEOREM

Quantum mechanics is the science of elementary particles. Spin is a property of elementary particles. For example, electrons are spin ½ particles and photons are spin 1 particles. When measuring the component of a spin 1 particle in a certain direction, the result will always be 0 or 1 or -1. And finally, when measuring the squared component of a spin 1 particle in three orthogonal directions, the result will always be (1,0,1) or (0,1,1) or (1,1,0). If I call the latter proposition the spin-axiom, then I can already formulate a version of the Kochen-Specker theorem.

Indeed, Kochen and Specker have shown that the spin-axiom implies that the spin-property of a spin 1 particle does not exist prior to spin-measurement.

The *reductio ad absurdum* proof of this theorem reads:

- Take a spin 1 particle and suppose its spin-components in all directions exist prior to any measurement; this is our supposition.

- Then the spin-axiom implies that for each triplet of orthogonal directions this particle has one pre-existing spin-squared-component that is 0 and two that are 1; this is a consequence of our supposition.

- But this consequence contradicts the two-color sphere theorem. Indeed, if I paint in green the intersection points of sphere S with each of its symmetry-axes belonging to a direction for which the pre-existing spin-squared-component is 0, and if I paint in red the intersection points of S with each of its symmetry-axes belonging to a direction for which the pre-existing spin-squared-component is 1, then the consequence implies that I can do exactly what the two-color sphere theorem shows to be impossible; this is our contradiction. Q.E.D.

As a metaphor to better understand the meaning of the Kochen-Specker theorem, think of a Poire Williams bottle.

The spin-components in all directions of a spin 1 particle cannot exist prior to any measurement, or it would mean I could do the impossible according to the two-color sphere theorem. Likewise, the size-components in all directions of the pear in the Poire Williams bottle cannot have existed prior to bottling, or it would mean someone could do the impossible, i.e., push a full-grown pear through a small bottleneck without crushing it. The spin property of an elementary particle is not a pre-existing property but emerges from the measurement process. Likewise, the size of a full-grown pear in a Poire Williams bottle is not a pre-existing property but emerges from a growth process as shown in the next picture.

In a sense, the circle has been closed. I started by highlighting that Whitehead conceives of beauty as primarily a property that emerges in the process of becoming of an occasion of experience. I ended by highlighting that quantum mechanics conceives of spin as a property that emerges in the process of measurement of an elementary particle.[1] All this invites us to further explore Whitehead's philosophy in which similar psychology-physics parallelisms constitute the starting point for his important metaphysical speculations.

But instead of continuing this chapter by joining Whitehead on his imaginative path, I end the chapter by reiterating the hope I uttered before. I hope that by highlighting the important role of the two-color sphere theorem in the wonderful worlds of particle physics and metaphysics, the reader has experienced agreement with the claim that the beauty of the two-color sphere theorem is best observed by presenting it as a crown jewel within the context of a theory, in this case quantum mechanics, hence verifying Rota's aphorism.

## NOTES

1    The process of becoming of an occasion of experience is a nontemporal micro-process; the process of growth of a pear in bottle is a temporal macro-process; but the process of measurement of an elementary particle is a controversial process: is it temporal or nontemporal, and a macro-process or a micro-process?

## REFERENCES

Adams Farmer, P. "Beauty Can Save Us: Mozart and Whitehead." In *Replanting Ourselves in Beauty: Towards an Ecological Civilization,* eds. Jay McDaniel & Patricia Adams Farmer, 105–09. Anoka, MN: Process Century Press, 2015.

Conway, J. & S. Kochen. "The Strong Free Will Theorem." *Notices of the AMS* 56.2 (2009): 226–32.

Henning, B. "Re-centering Process Thought: Recovering Beauty in A. N. Whitehead's Late Work." In *Beyond Metaphysics: Explorations in Alfred North Whitehead's Late Thought*, eds. Roland Faber, Brian G. Henning, & Clinton Combs, 201–14. Amsterdam & New York: Rodopi, 2010.

Kather, R. "The Web of Life and the Constitution of Human Identity: Rethinking Nature as the Main Issue of Whitehead's Late Metaphysics." In *Beyond Metaphysics: Explorations in Alfred North Whitehead's Late Thought*, eds. Roland Faber, Brian G. Henning, & Clinton Combs, 181–97. Amsterdam & New York: Rodopi, 2010.

King, J. P. *The Art of Mathematics*. New York: Dover Publications, 1993/2006.

Le Lionnais, F. "Beauty in Mathematics." In *Great Currents of Mathematical Thought*, ed. F. Le Lionnais, 121–58. New York: Dover Publications, 1948/1971.

Nagra, J. "Conway's Proof of the Free Will Theorem." 2005, cf. https://www.cs.auckland.ac.nz/~jas/one/freewill-theorem.html

Rota, G.-C. "The Phenomenology of Mathematical Beauty." In *Indiscrete Thoughts*, 121–33. Boston-Basel-Berlin: Birkhäuser, 1997/2008.

# 6. The Complementary Faces of Mathematical Beauty

JEAN PAUL VAN BENDEGEM

RONNY DESMET

Since antiquity philosophers have stressed that beauty is all about the emergence of unity amid diversity, and of the perfection of the unified whole from the harmony of its various parts with one another, and with the whole. In a sense, Whitehead's aesthetics of experience results from the application of these antique accounts of beauty to the patterns formed by the feelings that grow together in our experience. For him, beauty qualifies our experience when it is the unification of a variety of contrasting feelings, that is, the concrescence of many feelings of which the individual emotional intensities promote the intensity of the whole emotional pattern, and vice versa. As shown in Chapter Five, Whitehead's aesthetics of experience can inform the aesthetics of mathematical theorems—Euler's formula, the two-color sphere theorem. A mathematical theorem is beautiful if it has the inherent capability for the production of beauty of experience in a person with the appropriate background.

But what if we turn our attention from theorems to proofs? And what if we turn our attention from mathematical products to the production of mathematics? To what extent are the antique accounts of beauty in terms of unity and harmony still relevant when we focus on mathematical proof and creation? And how informative are Whitehead's writings when we broaden the discussion of mathematical beauty to include the aesthetics of the mathematical practices of proving and creating? This chapter aims at throwing some light on these issues by bringing together elements from the research of both authors in their respective fields of expertise: the philosophy of mathematical practices, and Whitehead's process philosophy.

## ECONOMIC AESTHETICS

"The feeling," Whitehead wrote, "widespread among mathematicians, that some proofs are more beautiful than others, should excite the attention of philosophers" (MT 60). So let us explore the notion of a beautiful proof by means of two examples that require no specialized mathematical knowledge.

The first example is Euclid's proof of the theorem that there are infinitely many primes. A prime number or prime is an integer greater than 1, which is only divisible by 1 and itself. Thus 2, 3, and 37 are examples of primes. Primes are the material out of which all integers greater than 1 are built. Indeed, they can all be expressed as a unique multiplication of primes, which is called their "prime factorization." Thus $37 = 37$ (for each prime the prime factorization is trivial) and $660 = 2.3.3.37$ (a prime factor can occur twice or more). Now consider any finite list of primes, say $p_1, p_2, \ldots, p_n$. Let P be the product of all the primes in this list, and let Q be P plus 1. Thus $Q = (p_1.p_2. \ldots .p_n)+1$. Also, let q be one of the factors of the prime factorization of Q (possibly $q=Q$, namely if Q is prime). Clearly, q is a prime other than $p_1, p_2, \ldots, p_n$ because Q is divisible by q (since q is a prime factor of Q), whereas Q is not divisible by $p_1, p_2, \ldots, p_n$ (since it leaves the remainder 1 when divided by any one of these primes). This proves that for any finite list of primes there is a prime not on the list, and therefore there must be infinitely many primes.

The second example is Pythagoras's *reductio ad adsurdum* proof of the theorem that the square root of 2 is irrational, i.e., that it cannot be expressed as a fraction. Suppose that the square root of 2 can be expressed as a fraction, in other words, that the square root of 2 equals p/q where p and q are integers and have no common factor (since if they had we could remove it). From this hypothesis (by squaring) follows that $2=p^2/q^2$. Hence $p^2=2q^2$. It follows that $p^2$ is even (since $2q^2$ is divisible by 2), and therefore that p is even (since the square of an odd number is odd). If p is even then $p=2s$ for some integer s, and therefore $2q^2=p^2=(2s)^2=4s^2$. Hence $q^2=2s^2$. It follows that $q^2$ is even, and therefore that q is even. However, if both p and q are even, they have a common factor, namely 2. This contradicts our hypothesis, and hence the hypothesis is false and the theorem true.

Versions of these two examples are also given in Godfrey Harold Hardy's famous 1940 essay, *A Mathematician's Apology*, in which—by the way—no author is quoted more often than Whitehead. According to Whitehead, "mathematics is concerned with the investigation of patterns of connectedness" (AI 153) or, shorter, "mathematics is the study of pattern" (ESP 111). Likewise, Hardy writes: "A mathematician, like a painter and a poet, is a maker of patterns" (84). And he adds: "The mathematician's patterns, like the painter's or the poet's must be *beautiful*; the ideas, like the colors or the words, must fit together in a harmonious way. Beauty is the final test: there is no permanent place in the world for ugly mathematics" (85). Hardy also agrees with Whitehead that it is an error to represent the love and aesthetic appreciation of mathematics as "a monomania confined to a few eccentrics in each generation" (85; also SMW 20), and he gives his account of Euclid's and Pythagoras's proofs because he is quite sure that most educated readers will be sensitive to their aesthetic appeal, and will recognize their beauty.

According to Hardy, beauty and seriousness are the two criteria by which the mathematician's patterns should be judged (98), and as beauty depends on seriousness (90), he first looks at "serious" theorems. Note that when Hardy speaks of theorems, he includes their proofs. Hardy writes: "A 'serious' theorem is a theorem which contains 'significant' ideas, and . . . the qualities which make a mathematical idea significant . . . are . . . a certain generality and a certain depth" (103).

According to Hardy, generality involves, among other things, being "capable of considerable extension," being "typical of a whole class of theorems," and being "a constituent in many mathematical constructs"

(104). Take Pythagoras' proof of the irrationality of the square root of 2. It can be extended to the square root of 3, namely, by replacing "even" (which means: "being a multiple of 2") by "being a multiple of 3," and by noticing that if the square of an integer is a multiple of 3, so is the integer itself (since if the integer itself does not have 3 as a factor in its unique prime factorization, its square cannot have it as such a factor either). Pythagoras' reasoning is typical of a whole class of proofs, which includes the proofs of the irrationality of the square root of 2, of 3, of 5, of 7, and so on. In other words, it can be generalized to the proof of the irrationality of the square root of any prime. Moreover, the proof that the square root of each of the primes cannot be a fraction implies the construction of a whole class of non-fractions or irrational numbers to supplement the fractions or rational numbers. This is another element contributing to the generality (and, hence, the seriousness or significance or importance) of Pythagoras' irrationality proof, and it is also an element contributing to the depth of this proof. Indeed, Hardy holds that "the idea of an 'irrational' is deeper than that of an integer" (or a fraction of integers), and that "Pythagoras' theorem is, for that reason, deeper than Euclid's" (110). To Hardy, "it seems that mathematical ideas are arranged somehow in strata" (110), and he imagines the world of numbers as consisting of ever deeper strata, and the stratum of the irrational numbers as deeper than that of the integers. Moreover, Hardy has another reason to think that Euclid's proof, even though very important, is not very deep: depth also involves "using the most powerful weapons" of mathematics, but "we can prove that there are infinitely many primes without using any notion deeper than that of 'divisibility'" (111). However, as Hardy admits, "this notion of 'depth' is an elusive one even for a mathematician who can recognize it" (112).

After dealing with the seriousness of proofs, Hardy turns to their aesthetic qualities, and writes:

> What 'purely aesthetic qualities' can we distinguish in such theorems as Euclid's and Pythagoras'? I will not risk more than a few disjointed remarks.
>
> In both theorems (and in the theorems, of course, I include the proofs) there is a high degree of *unexpectedness*, combined

with *inevitability* and *economy*. The arguments take so odd and surprising a form; the weapons used seem so childishly simple when compared to the far-reaching results; but there is no escape from the conclusions. There are no complications of detail—one line of attack is enough in each case; and this is true too of the proofs of many much more difficult theorems, the full appreciation of which demands quite a high degree of technical proficiency. We do not want many 'variations' in the proof of a mathematical theorem: 'enumeration of cases,' indeed, is one of the duller forms of mathematical argument. A mathematical proof should resemble a simple and clear-cut constellation, not a scattered cluster in the Milky Way. (113)

This quote invites us to make three important comments.

(1) One might argue against unexpectedness or surprise as an essential aesthetic quality by giving many cases of beautiful proofs, since for some of them the quality of surprise is absent, whereas for others it is present and adds to the experience of beauty. And indeed, the search for the essential characteristics of beauty seems to be as idle as Ludwig Wittgenstein's search for the essential characteristics of games. This observation can inspire to give up the search for the essence of mathematical beauty, and to introduce, in Wittgenstein's footsteps, the notion of family resemblance in order to give a non-essentialist, yet coherent account of beautiful proofs. Here, however, we focus on Whitehead not Wittgenstein.

(2) One might argue that all proofs are inevitable and, hence, that inevitability, even though it contributes to the beauty of proofs, does not shed any light on why one proof is more beautiful than another. And indeed, to the extent that only logically ideal proofs are taken to be real proofs, they are all inevitable. This remark presupposes that agreement rules with respect to the logical ideal, whereas the definition of this ideal is the topic of heated debates. Here, however, we ignore these debates.

(3) Hardy does not think that a case-by-case proof has the kind of simplicity that contributes to a mathematician's aesthetic delight. For example, the proof of the two-color sphere theorem given in Chapter Five "is just that 'proof by enumeration of cases' (and of cases which do not, at bottom, differ at all profoundly) which," according to Hardy, "a

real mathematician tends to despise" (114).[1] Also, Hardy does not equate economy with the right kind of simplicity as such, but conceives of a proof as economic if it is simple *in comparison with* its seriousness (its significance, its importance, its being far-reaching). Here, we take Hardy's aesthetic quality of economy, which explains how beauty depends on seriousness, to be the most relevant quality for further discussion.

In Henri Poincaré's 1908 book, *Science and Method*, we can witness a similar interplay of beauty, importance, generality, surprise, economy, simplicity, and so on:

> What is it that gives us the feeling of elegance in a solution or a demonstration? It is the harmony of the different parts, their symmetry, and their happy adjustment; it is, in a word, all that introduces order, all that gives them unity, that enables us to obtain a clear comprehension of the whole as well as of the parts. But that is also precisely what causes it to give a large return; and in fact the more we see this whole clearly and at a single glance, the better we shall perceive analogies with other neighboring objects, and consequently the better chance we shall have of guessing the possible generalizations. Elegance may result from the feeling of surprise caused by the unlooked-for occurrence together of objects not habitually associated. In this, again, it is fruitful, since it thus discloses relations till then unrecognized. It is also fruitful even when it only results from the contrast between the simplicity of the means and the complexity of the problem presented, for it then causes us to reflect on the reason for this contrast, and generally shows us that this reason is not chance, but is to be found in some unsuspected law. Briefly stated, the sentiment of mathematical elegance is nothing but the satisfaction due to some conformity between the solution we wish to discover and the necessities of our mind, and it is on account of this very conformity that the solution can be an instrument for us. This aesthetic satisfaction is consequently connected with the economy of thought. (375–76)

Like Hardy, Poincaré connects aesthetic satisfaction with economy. He considers "economy of thought" to be "a source of beauty" (369), and he links it to importance by saying that "the importance of a fact is measured by the return it gives — that is, by the amount of thought it enables us to economize" (374). Applied to "mathematical demonstration," Poincaré holds that the return of a proof involves the "feeling" or "intuition" of "the whole of the argument at a glance" (389–90). Moreover, an economic and important proof, in which we see "all parts in a single glance" (376), allows us to "summarize it in a few lines" for "those that come after us" (377), and to "perceive immediately what must be changed to adapt it to all the problems of a similar nature that may be presented" (376). Initially, finding a solution to a particular problem, and proving it, can involve "blind groping," but to this Poincaré adds:

> My time will not have been lost if this very groping has succeeded in revealing to me the profound analogy between the problem just dealt with and a much more extensive class of other problems; if it has shown me at once their resemblances and their differences; if, in a word, it has enabled me to perceive the possibility of a generalization. Then it will not be merely a new result that I have acquired, but a new force. (374)

Poincaré's account is *similar* to Hardy's account, not identical. So there *are* differences. The main one is that Poincaré, contrary to Hardy, explicitly links surprise, economy, simplicity, importance, and generality with the human mind, with the amount of thought, with perception and intuition of the whole at a single glance, with the return of groping and reasoning, and with the recognition of analogy. In other words, Poincaré links beauty to the *subject* that feels, perceives, intuits, gropes, reasons, and recognizes, whereas Hardy aims at a more *objective* aesthetic account in terms of patterns of *ideas*. Poincaré's account is Kantian, whereas Hardy's is Platonic.[2] For Poincaré, in the field of mathematical activity, a person's interior forms of intuition crystalize in mathematical patterns,[3] whereas for Hardy mathematical patterns are simply a person's notes of observations of the exterior reality of mathematics. This also helps to explain the quite amazing fact that Hardy hardly talks of the aesthetic role that is played by self-evidence, understanding, insight, and enlightenment. However, despite the difference of philosophical language, Hardy's and Poincaré's accounts of mathematical beauty both culminate in a discussion of a similar economic aesthetics, and so does George David Birkhoff's search for an "aesthetic measure" in his 1933 book, *Aesthetic Measure*.

In *Aesthetic Measure*, Birkhoff proposes a formula to quantify the qualitative, namely, to measure the aesthetic value as experienced by art-lovers and mathematicians alike. Birkhoff's analysis of the aesthetic experience suggests that the aesthetic measure $M$ equals the harmony, symmetry, or order $O$ of the artistic or mathematical object, divided by the complexity $C$, which is proportional to the preliminary effort necessary for the act of perception (3–4). Birkhoff's general comment on the formula $M=O/C$ reads: "The well known demand for 'unity in variety' is evidently closely connected with this formula" (4) since "it is the intuitive estimate of the amount of order $O$ inherent in the aesthetic object, as compared with its complexity $C$, from which arises the derivative feeling of the aesthetic measure $M$" (11–12).

A first way of reading Birkhoff's formula, $M=O/C$, is the following: when the order in an object increases, it will be experienced as more beautiful; when the complexity increases, it will be experienced as more ugly. This way of reading and applying the formula leads to the confirmation of commonplaces such as the antique cliché that the circle is the acme of beauty in mathematics because of its infinite symmetry and utter simplicity; or the prejudice that the most beautiful music is Western music based on the classical scales and harmony; or the banality that in all arts beauty is increased by symmetry. But then what are we to do with Piet Mondrian's explicit rejection of symmetry for being a negative property? Does this imply that all his paintings are ugly? And do we have to conclude from Birkhoff's formula that Arnold Schönberg, Alban Berg, and Anton Webern have heralded the end of beautiful music? And what about the contemporary mathematician's experience of the complex beauty of fractals, which defies the simplistic beauty of the circle? Clearly, Birkhoff's formula cannot deliver on the promise of being an adequate measure of the value of our aesthetic experiences. Rather, it lays bare the problematic nature of all quantitative aesthetics. This is not, however, our final verdict on Birkhoff's approach.

Another and more interesting way of reading Birkhoff's formula is suggested by two analogies that he offers to justify it. He writes: "The definition of the beautiful as that which gives us the greatest number of ideas in the shortest space of time (formulated by Heemsterhuis in the eighteenth century) is of an analogical nature" (4). And Birkhoff draws a second analogy "from the economic field":

> In each business there is involved a certain investment $i$ and a certain annual profit $p$. The ratio $p/i$, which represents the percentage of interest on the investment, is regarded as the economic measure of success. Similarly in the perception of aesthetic objects . . . there is involved a feeling of effort of attention, measured by $C$, which is rewarded by a certain tone of feeling, measured by $O$ . . . By analogy, then, it is the ratio $O/C$ which best represents the aesthetic measure $M$. (12)

These two analogies suggest a second interpretation of $M=O/C$, which is in line with Hardy's and Poincaré's economic accounts of mathematical beauty, especially when applied to a mathematical proof $P$. Indeed: $O(P)$ can be taken as directly proportional to Hardy's qualities of depth and generality, and $C(P)$ as inversely proportional to his quality of simplicity. Alternatively, $O(P)$ can be taken as directly proportional to Poincaré's intuition of the whole and the possibility of generalization, and $C(P)$ as directly proportional to Poincaré's amount of thought needed. Synthesizing, we can take $O(P)$ as a measure of the mathematician's intuition of the depth and generality of the whole proof, and $C(P)$ as a measure of the mathematician's time and effort to overcome the difficulty of the proof, that is, to unify the variety of mathematical ideas and logical deductions constituting the proof. Consequently, and analogous to the return-on-investment formula in economy, the aesthetic measure formula $M(P)=O(P)/C(P)$ then means that the aesthetic satisfaction produced by a proof can be defined as the return of intuition of mathematical depth and generality of the whole on investment of time and effort in proof.

This second, economic or return-on-investment interpretation of Birkhoff's formula, inspired by Hardy's and Poincaré's economic aesthetics of mathematical proof, and by Birkhoff's two analogies, is not only the interpretation we favor, but also a remarkably good summary of Whitehead's aesthetics of mathematical proof. Indeed, for Whitehead too, the beauty of a proof is all about the return of holistic or large-scale intuition on investment of logical effort in this proof. In *Modes of Thought*, Whitehead utilizes "self-evidence" and "understanding" as synonyms for "intuition," and he writes:

> There is very little large-scale understanding, even among mathematicians. There are snippets of understanding, and snippets of connections between these snippets. These details of connection are also understood. But these fragments of intelligence succeed each other. They do not stand together as one large self-evident coordination. At the best, there is a vague memory of details which have recently been attended to.
>
> This succession of details of self-evidence is termed *proof*. But the large self-evidence of mathematical science is denied to humans. (MT 46–47)

And then, after emphasizing "the great variety of characters that self-evidence can assume, both as to extent and as to the character of the compositions which are self-evident" (MT 47),[4] Whitehead continues:

> The thesis that I am developing conceives proof, in the strict sense of that term, as a feeble second-rate procedure. When the word *proof* has been uttered, the next notion to enter the mind is halfheartedness. Unless proof has produced self-evidence and thereby rendered itself unnecessary, it has issued in a second-rate state of mind, producing action devoid of understanding. Self-evidence is the basic fact on which all greatness supports itself. But proof is one of the routes by which self-evidence is obtained. (MT 48)

A consequence of this doctrine, Whitehead claims, is that "proof should be at a minimum. The whole effort should be to display . . . self-evidence" (MT 48). On the other hand, Whitehead adds:

> Our understanding is not primarily based on inference. Understanding is self-evidence. But our clarity of intuition is limited, and it flickers. Thus inference enters as a means for the

attainment of such understanding as we can achieve. Proofs are the tools for the extension of our imperfect self-evidence. (MT 50)

Whitehead's next claim is that the return of intuition or self-evidence (or insight or understanding) on investment of logical reasoning in formal proof is not merely a cognitive return, but an affective return as well, namely, that it involves not only proof-induced enlightenment, but also aesthetic delight. Prior to giving a long Whitehead quote justifying this claim, we first highlight that it is a controversial one.

In "The Phenomenology of Mathematical Beauty," starting in conformity with Whitehead's view but ending with a critique of the economic aesthetics of mathematical beauty, Gian-Carlo Rota writes:

> Every teacher of mathematics knows that students will not learn by merely grasping the formal truth of a statement. Students must be given some enlightenment . . . or they will quit.[5]

> If the statements of mathematics were formally true but in no way enlightening, mathematics would be a curious game played by weird people. Enlightenment is what keeps the mathematical enterprise alive.

> Mathematicians seldom explicitly acknowledge the phenomenon of enlightenment for at least two reasons. First, . . . enlightenment is not easily formalized. Second, enlightenment admits degrees: some statements are more enlightening than others. Mathematicians dislike concepts admitting degrees . . . Mathematical beauty is the expression mathematicians have invented in order to obliquely admit the phenomenon of enlightenment while avoiding acknowledgment of the fuzziness of this phenomenon. They say that a theorem is beautiful when they mean to say that the theorem is enlightening. . . . The term "mathematical beauty" . . . is a trick mathematicians have devised to avoid facing up the messy phenomenon of enlightenment. (132)

Whitehead does not fail to face the phenomenon of enlightenment by degree, nor to acknowledge the vagueness of the notion of enlightenment, and of the similar notions of intuition, self-evidence, understanding, and insight. But it is true that, like most mathematicians, he holds that a proof is beautiful when the return of enlightenment on investment in proof is high. So the question arises whether Whitehead's correlation of aesthetic delight with proof-induced self-evidence is indeed a trick that philosophers better avoid. The answer to this question is "No," because such a correlation is only a fallacious trick in a philosophy which separates the faculty of affection from that of cognition, and which posits the correlation without justification— Whitehead's philosophy does neither of these two. As highlighted in Chapter Five, Whitehead rejects all faculty psychology, and, as we will highlight now, he justifies the intimate relation between the aesthetic delight in a proof and the logic-induced enlightenment it involves.

According to Whitehead, "the feeling, widespread among mathematicians, that some proofs are more beautiful than others" can be justified by the fact that *"aesthetic experience is another mode of the enjoyment of self-evidence"* (our italics), and he adds:

> I suggest to you that the analogy between aesthetics and logic is one of the undeveloped topics of philosophy.
>
> In the first place, they are both concerned with the enjoyment of a composition, as derived from the interconnection of its factors. There is one whole, arising from the interplay of the many details. The importance arises from the vivid grasp of the interdependence of the one and the many. If either side of this antithesis sinks into the background, there is trivialization of experience, logical and aesthetical.
>
> The distinction between logic and aesthetics consists in the degree of abstraction involved. Logic concentrates upon high abstraction, and aesthetics keeps . . . close to the concrete . . .
>
> The characteristic attitude of logical understanding is to start with the details, and to pass to the construction achieved. Logical enjoyment passes from the many to the one. . . . The understanding of logic is the enjoyment of the abstracted details as permitting that abstract unity. As the enjoyment develops, the revelation is the unity of the construct.

> The movement of aesthetic enjoyment is in the opposite direction. We are overwhelmed with the beauty of the build- ing . . . The whole precedes the details. We then pass to dis- crimination. (MT 61–62)

In other words, according to Whitehead it is justified to correlate aesthetic enjoyment (aesthetic delight) with logical enjoyment (logic- induced enlightenment) because the enjoyment of mathematical beauty is a mode of the enjoyment of self-evidence. But whereas logical enjoyment develops from the variety of details to their unification in an abstract pattern, the aesthetic enjoyment is an emotional flow from the concrete whole to the discrimination of details.

We could end the first part of our chapter here, but instead we choose to extend the above quote a little. The reason for this is that what Whitehead writes next forms an ideal introduction to the second part. First Whitehead addresses the doubt whether or not, when the topic of aesthetic enjoyment is sufficiently explored, there will be anything left for discussion:

> This doubt is unjustified. For the essence of great experience is penetration into the unknown, the unexperienced. . . . Our lives are passed in the experience of disclosure. As we lose this sense of disclosure, we are . . . descending to mere conformity with . . . the past. Complete conformity means the loss of life. (MT 62)

> The essence of life is to be found in the frustrations of established order. The Universe refuses the deadening influence of complete conformity. And yet in its refusal, it passes towards novel order as a primary requisite for important experience. (MT 88)

Frustrating the established aesthetic canon to enable the emergence of a novel one is characteristic of great artists and mathematicians alike. Think, for example, of David Bowie, a musician and performer who, again and again, renewed his music and performance. Bowie died while we were writing the first part of this chapter, and in one of the many interviews rebroadcasted following his death, Bowie said that whenever he started to see a thread of stability in what he was making, he felt the impulse of destroying it, and he added: "When you can predict the outcome of what you are doing, that's incredibly unsatisfactory." Bowie did not stop being a non-conformist and feeling the urge to make new kinds of songs and acts. He will be remembered as forever young in accord with what Whitehead once wrote: "Youth is not defined in years but by the creative impulse to make something" (AE 119). And what holds for Bowie, holds for great mathematicians too, which is why Eric Temple Bell once wrote: "The essence of mathematics is its eternal youth" (quoted by Le Lionnais 144). The creative and youthful character of mathematics is the topic of the second part of this chapter.

## FROM PRODUCT AESTHETICS TO CREATION AESTHETICS

When we think of the beauty of theorems and proofs, our thinking is primarily product-focused. It is the end result of the mathematical activity that is being evaluated aesthetically. That is why it is appropriate to say that our account so far has dealt with product aesthetics. However, the exploration of new territory is as important a part of mathematics as the enjoyment of old territory. In fact, the history of mathematics is the story of the exploration of new territory. The explorations of analytic geometry, group theory, non-Euclidean geometry, complex analysis, and chaos theory, are but a few examples of important historical breakthroughs in mathematics.

The exploration of fractals, one of the relatively recent branches of the tree of mathematical research, is most appropriate as an illustration to introduce the 1948 essay of François Le Lionnais, "Beauty in Mathematics," which will guide our thinking beyond a mere product aesthetics towards a creation aesthetics. When asking mathematicians what is so beautiful about fractals, one is likely to receive answers such as: "because they are so intricate," or: "because an extremely simple reiteration formula can generate an extremely complex object," or: "because no one expected such complexity to be hidden in mathematics." These answers feel bizarre because they seem to flatly contradict Birkhoff's view that mathematical beauty is about minimizing complexity rather than generating it. With respect to the exploration of fractals, however, mathematicians seem to behave like "romantics" — this is Lionnais's term, and it stands for his fundamental insight with respect to mathematical exploration, research, and creation.

"Works of art can," according to Le Lionnais, "be ranged under two grand banners: *classicism*, all elegant sobriety, and *romanticism*, delighting in striking effects and aspiring to passion" (123), and he holds that it is also possible to distinguish classical beauty and romantic beauty in mathematics.

"We say," Le Lionnais writes, "that a mathematical proposition has classical beauty when we are impressed by its austerity or its mastery over diversity, and even more so when it combines these two characteristics in a harmoniously arranged structure" (124). Clearly, according to Le Lionnais, classical beauty involves unity amid diversity, and harmony of parts and whole; and what he writes next on classical beauty can be read as a repetition of our account of Hardy and Poincaré. Classical beauty depends on the "value" or importance of a mathematical theorem, which, in its turn, "depends on the depth of the mathematics required to prove it" (126) as well as on its generality, that is, the light it throws on the "sublime interdependence" of mathematical concepts, theorems, proofs, and theories (130): "Classic are the methods which cast a new light on previously known facts, bringing together and unifying discoveries formerly considered disparate" (137). Moreover, classical beauty involves unexpectedness, for it "intrigues us especially when we are expecting a certain disorder" (124), as well as economy: "It seems to us that a method earns the epithet of classic when it permits the attainment of powerful effects by moderate means" (136).

"By contrast with classical mathematical beauty," Le Lionnais writes, "we are now going to examine another sort of beauty which can be described as romantic. Its underlying principle is the glorification of violent emotion, non-conformism and eccentricity" (130). Le Lionnais first states that when mathematicians behave like romantics (say, like Bowie), that is, when they are in exploration-and-creation mode, their violent rejection of conformism can produce "what seem to be completely illogical results repugnant to common sense" (132). And then Le Lionnais gives a spectrum of examples, from which we select the example of Georg Cantor's exploration of transfinite set theory, and the example of the exploration of continuous functions without derivatives. With respect to the first of these two examples, Le Lionnais writes:

> Does not the modern theory of sets take as its point of departure concepts which seemed an insolent defiance of common sense when Cantor defended them? This exuberant theory had to enjoy repeated successes in other disciplines already classic like arithmetic and analysis before we would accept the existence of quantities "greater than infinity" (Cantor's expression) plus the startling number $\omega$ situated *on the other side of infinity.* Theologians were not the last to protest certain ideas as unfair competition.

> After the paradoxes come the anomalies, the irregularities, indeed the monstrosities. They arouse some people's indignation and to others bring delight. (133)

For example, Cantor's romantic exploration of transfinite set theory aroused Leopold Kronecker's indignation, and Kronecker famously wrote: "I do not know what predominates in Cantor's theory — philosophy or theology, but I am sure that there is no mathematics there." On the other hand, it delighted David Hilbert, who famously wrote: "No one will drive us from the paradise which Cantor created for us."[6] Today, the romantic exploration of Cantor has been turned into a more classical enjoyment of transfinite set theory, and most mathematicians can share Hilbert's aesthetic delight. This is the way mathematics advances: once a new territory has been mapped out, and its problems have been identified, the classicist ideal takes over again — problems must be solved, proofs must be found, and the investment of effort in logical reasoning must give a significant return of self-evidence.

With respect to the second of the examples we selected, some preliminary remarks are due. A continuous function is one of which you can draw the graph without lifting your pen. And a continuous function has a derivative in a point of its graph, if it is possible to draw a tangent to this graph through this point. For a long time it was intuitively assumed that all continuous functions have derivatives in all points of their graphs, except in a few, namely, those points in which these graphs are changing their direction not smoothly, but abruptly.

This intuitive assumption, however, was falsified by the romantics who explored continuous functions *without* derivatives, that is, of continuous graphs (and curves in general) that are *nowhere* smooth and, hence, do not have a single tangent. With respect to these romantics, Le Lionnais writes:

> When Riemann and Weierstrass made known the existence of continuous functions without derivatives, what an outcry came from the mathematicians against these newcomers: "I turn with fright and horror from this lamentable plague of continuous functions having no derivatives," exclaimed Charles Hermite. If it is difficult to reason about such functions, it becomes impossible to visualize fully the infinite caprices of the curves representing them. (134)

Not only Hermite was horrified—Poincaré too! And in *Science and Method* Poincaré called these "weird" continuous functions without derivatives, "monsters," and the ensemble of curves without tangents, a "collection of monstrosities" (447). Consider, however, Helge von Koch's construction of the curve without tangents that is called the Koch snowflake: start with an equilateral triangle, and then recursively alter each line segment in three steps: (1) divide the line segment in three segments of equal length; (2) draw an equilateral triangle that has the middle segment of step 1 as its base and points outward; (3) remove the line segment that is the base of the triangle of step 2. In the figure below you can see the initial triangle and the first five iterations, but the Koch snowflake itself is the limit approached as the above steps are followed over and over again, and so it cannot be pictured.

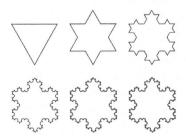

When the Koch snowflake was discovered-cum-invented in 1904, it was indeed an example of a curve without tangents, that is, one of Poincaré's "monstrosities." However, today, the Koch snowflake is known as one of the many examples of beautiful fractals, and Le Lionnais writes:

> The romantic wildness of continuous functions without derivatives could evoke in the mystical Hermite the impression that he was battling demons escaped from some mathematical hell. Observe, however, the case of one of these functions, the celebrated Koch curve or homunculus. Every arc of this curve, no matter how short, is similar to the entire curve, whose exquisite arabesque it chisels into infinity with unfailing regularity. What could be more classical? (136)

Today, the Koch snowflake can be classified among other fractals by using a generalized notion of dimension. A straight line still has dimension 1, and the plane still has dimension 2, but curves like the Koch snowflake are given a dimension in between 1 and 2, $\log4/\log3$ to be exact. In other words, the Koch snowflake gradually became part of a larger whole, and the romantic exploration of the incoherent and wild territory of curves without tangents gradually turned into a classical enjoyment of the ever more coherent and structured territory of fractals. Today, we might say, Poincaré's monsters have been tamed. Clearly, once the more romantic mathematicians have explored and mapped out the new territory, they hand over their maps to more classical mathematicians for further refinement, and there is beauty to be found in both the romantic explorations and the classical refinements.

"We must not think," Le Lionnais concludes, "that mathematics progresses only via the royal road of classicism" (138). Mathematics is in need of both classicism and romanticism, of both classical and romantic mathematicians. In fact, mathematical progress consists essentially in "an ever-renewed antagonism" between the classical "desire for unity" and the romantic "rebellions" which erupt at every new attempt to get at the heart of mathematics (141). So if we want to arrive at a satisfactory aesthetics of mathematics, we cannot separate mathematics from other art forms, but will have to take into account at least the two basic forms here treated:

the classical or economic aesthetics that is primarily oriented on the (end) products of mathematics, especially proofs; and the romantic or exploring aesthetics that is oriented on prospection of new mathematical territory.

Le Lionnais's essay makes clear the role of romantic exploration in the progress of mathematics, but it does not offer a clear definition of romantic beauty as opposed to classical beauty. We learn that romantic beauty involves striking effects and violent emotions, but do not get anywhere near a clear-cut definition such as the antique unity and harmony definitions of classical beauty, or their Whiteheadian offspring. In Chapter Five, Farmer's evocation of Mozart's Jupiter Symphony sheds light on classical beauty in mathematics. Similarly, in *Emblems of Mind: The Inner Life of Music and Mathematics,* Edward Rothstein's invocation of Chopin's Prelude in A major sheds light on romantic beauty in mathematics (180–189). Skipping all the details he gives, and immediately focusing on his conclusion, Rothstein holds that whereas classical beauty enlightens and delights, romantic beauty disturbs and overwhelms, and, following Immanuel Kant, Rothstein reserves the term "beautiful" to qualify an instance of classical beauty, and the term "sublime" to qualify an instance of romantic beauty. In fact, the beautiful and the sublime are so different, that one might question the use of the expression "romantic beauty." And what about Whitehead? Whitehead's writings do not contain a theory of romantic beauty or, better, of the sublime. And yet, the romantic spirit, which Le Lionnais claims to be vital to mathematical progress, is prominently present in Whitehead's oeuvre. This youthful spirit of creativity, which sacrifices aesthetic harmony and delight to be rewarded with it at a higher level, and which Ferdinand Gonseth called "the spirit of adventure" (quoted by Le Lionnais 144), is a key ingredient of the aesthetics of experience that Whitehead develops, for example, in his *Adventures of Ideas.*

In *Adventures of Ideas*—a romantic title indeed—Whitehead highlights that there are different types of beauty (AI 252), and that it is important, when developing a theory of aesthetics, not only to distinguish lower and higher types of beauty, but also to account for the role of discordant feelings in the transition from type to type. According to Whitehead:

There are in fact higher and lower perfections, and an imperfection aiming at a higher type stands above lower perfections. . . . Progress is founded upon the experience of discordant
feelings. The social value of liberty lies in its production of dis-
cords. There are perfections beyond perfections. All realization
is finite, and there is no perfection which is the infinitude of
all perfections. Perfections of diverse types are among themselves discordant. Thus the contribution to Beauty which can
be supplied by Discord — in itself destructive and evil — is the
positive feeling of a quick shift of aim from the tameness of
outworn perfection to some other ideal with its freshness still
upon it. . . .

Ancient Greek civilization . . . attained its proper beauty in
human lives to an extent not surpassed before or since. Its arts,
it theoretic sciences, its modes of life, its literature, its philosophic schools, its religious rituals, all conspired to express every
aspect of this wonderful ideal. Perfection was attained, [but]
with the attainment inspiration withered. With repetition in
successive generations, freshness gradually vanished. Learning
and learned taste replaced the ardor of adventure. . . . To sus-
tain a civilization with the intensity of its first ardor requires
more than learning. Adventure is essential, namely, the search
for new perfections. (AI 257–58)

Whitehead does not explicitly deal with mathematics here, but it is
obvious that what he says can be applied to mathematics. The destructive feelings that are needed when a civilization falls prey to the tameness of outworn perfection are none other than the romantic feelings
that are needed when a mathematical theory "which overwhelms us
the first time we meet it, comes to appear trite in the end" (Le Lionnais
127); the discordant feelings that lie at the basis of human progress in
general are none other than the sublime feelings that lie at the basis of
mathematical progress in particular; and instead of giving the example
of Greek civilization, Whitehead could just as well have focused on the
example of Greek mathematics.

Whitehead claims that adventure — the search for new perfections — is essential to the advance of humanity. In the context of

mathematics, this claim is not different from Le Lionnais' claim that romanticism — the search for novel aesthetic delights — is essential to the advance of mathematics.

In *Process and Reality*, Whitehead writes:

> There are various types of order, and some of them provide more trivial satisfaction than do others. Thus, if there is to be progress beyond limited ideals, the course of history by way of escape must venture along the borders of chaos in its substitution of higher for lower types of order. (PR 111)

Very few additions are needed to turn this quote into an appropriate ending of our discussion of the romantic aesthetics of mathematical creation:

> There are various types of order, and some of them provide more trivial aesthetic satisfaction than do others. Thus, if there is to be progress beyond limited mathematical ideals, the course of the history of mathematics must venture along the borders of chaos in its substitution of higher for lower types of order.

## CONCLUSION

Whitehead does not have a separate classical aesthetics of mathematical proof, but his aesthetics of experience can account for the aesthetic delight in proof, which emerges from the return of intuitive enlightenment on investment of logical effort in proof. Also, Whitehead does not have a separate romantic aesthetics of mathematical creation, but his aesthetics of experience can account for the romantic and adventurous explorations of the mathematical landscape, which lead mathematicians from one mountain top to another, from one panoramic view to another, from one type of order to another, and from aesthetic delight to aesthetic delight, but not without descending in the valleys of discord, not without suffering the loneliness of non-conformism, not without venturing along the borders of chaos, and not without exercising the challenging freedom that can save us from our smallness and lack of vision.

Whitehead's aesthetics of experience, even though developed *by* a mathematician, is not explicitly focused *on* mathematics. And yet, it

allows us to understand not only that it is "alluring aesthetic satisfac-
tions which have motivated modern mathematicians to cultivate their
cherished study with such ardor" (Le Lionnais 121), but also that it is
"the spirit of adventure (which) animates the mathematician far more
than his formulas" (Gonseth, quoted by Le Lionnais 144), or, in other
words, that Cantor is right: "The essence of mathematics lies precisely
in its freedom" (896).

## NOTES

1    A case-by-case proof often proves a theorem to be true without shedding
     light on *why* it is true, and less enlightenment implies less aesthetic delight.
     For example: the proof of the two-color sphere theorem does not shed light
     on *why* its cases culminate in the impossible task of having to paint one
     point in two colors; it is quite ugly.

2    As is clear from Chapter Five, Whitehead's aesthetics of experience
     emphasizes both the subjective *and* the objective character of beauty.
     Also, even though Whitehead adopts some philosophical ideas of Kant,
     and several of Plato, his philosophy is neither Kantian nor Platonic.

3    Poincaré's forms of intuition, however, do not involve Kant's sensible
     intuition of space and time, but the intellectual intuition of mathematical
     induction, continuity, and groups.

4    For example, Ramanujan, contrary to Whitehead, had an impressively
     extensive intuition with respect to numerical patterns; and Whitehead,
     contrary to Ramanujan, especially enjoyed "patterns of relationships in
     which numerical and quantitative relationships are wholly subordinate"
     (MT 47).

5    In his educational writings, Whitehead conceives of education as a cyclic
     process in which each cycle consists of three stages: first the stage of
     romance, then the stage of precision, and finally, the stage of generaliza-
     tion. These stages, continually recurring in cycles, determine what White-
     head calls "the rhythm of education" (AE 15). Roughly, and applied to
     mathematics, on might say that Whitehead's three stages are the stage of
     undisciplined intuition, the stage of logical reasoning, and the stage of
     logically guided intuition. By skipping stage one, and never arriving at
     stage three, bad math teachers deny students the only possible motiva-
     tion to love mathematics: enlightenment! Unfortunately, as Paul Lock-
     hart correctly claims in *A Mathematician's Lament*, at present most math
     teachers are forced by the standard school mathematics curriculum to be
     bad math teachers: "What is happening is the systematic undermining

of the student's intuition," whereas the goal of each proof, that is, of each mathematical argument, is to be "a beacon of light – it should refresh the spirit and illuminate the mind" (68).

6   Less known than Hilbert's remark in "Über das Unendliche," *Mathematische Annalen* 95(1926): 367–94, is the fact that Whitehead, who wrote very few mathematical research papers, nonetheless published three papers in the *American Journal of Mathematics*, which dealt with Cantor's theory in the context of Peano's developments of mathematical logic, and Russell's symbolism for the logic of relations: "On Cardinal Numbers" *AJoM* 24(1902): 367–94; "The Logic of Relations, Logical Substitution Groups, and Cardinal Numbers" *AJoM* 25(1903): 157–78; "Theorems on Cardinal Numbers" *AJoM* 26(1904): 31–32.

## REFERENCES:

Birkhoff, G. D. *Aesthetic Measure*. Cambridge, MA: Harvard University Press, 1933.

Cantor, G. "Foundations of a General Theory of Manifolds: A Mathematico-Philosophical Investigation into the Theory of the Infinite." In *From Kant to Hilbert: A Source Book in the Foundations of Mathematics — Volume II*, edited by Wiliam Ewald, 878–920. Oxford: Clarendon Press, 1883/2005.

Hardy, G. H. *A Mathematician's Apology*. Cambridge: Canto Edition, 1940/1993.

Le Lionnais, F. "Beauty in Mathematics." In *Great Currents of Mathematical Thought*, edited by François Le Lionnais, 121–158. New York: Dover Publications, 1948/1971.

Lockhart, P. *A Mathematician's Lament*. New York: Bellevue Literary Press, 2009.

Poincaré, H. *Science and Method*. In *The Value of Science: Essential Writings of Henri Poincaré*, edited by Stephen Jay Gould, 355–572. New York: Modern Library, 1908/2001.

Rota, G.-C. "The Phenomenology of Mathematical Beauty." In *Indiscrete Thoughts*, 121–133. Boston-Basel-Berlin: Birkhäuser, 1997/2008.

Rothstein, E. *Emblems of Mind: The Inner Life of Music and Mathematics*. Chicago and London: University of Chicago Press, 1995/2006.

# 7. Creating a New Mathematics

ARRAN GARE

In this chapter, my focus is on efforts to create a new mathematics, with my prime interest being the role of mathematics in comprehending a world consisting first and foremost of processes, and examining what developments in mathematics are required for this. I am particularly interested in developments in mathematics able to do justice to the reality of life. Such mathematics could provide the basis for advancing ecology, human ecology, and ecological economics and thereby assist in the transformation of society and civilization so that we augment life rather than undermining the conditions for our existence. It was in the process of grappling with these problems that I was drawn to investigate the tradition of intuitionism in mathematics and the role of intuition in mathematics, science, and philosophy, and then to consider Whitehead's work on mathematics and its philosophy in relation to these.

This is part of a broader project. As I see it, the defense of process philosophy is a struggle against the nihilism of European civilization, brought about, as Nietzsche argued, by the will to power turned against itself. While it is usual to interpret this in relation to Christian morality, it is clear from Nietzsche's *Philosophical Notebooks* that it was the transmogrification of this morality into the quest for scientific truth that most concerned him. It is this quest for truth that has led to the denial of reality to creative processes and life, since the reality of these stands in the way of comprehending the world as transparent to reason and thereby, predictable and controllable. As Nietzsche noted, "To impose upon becoming the character of being — that is the supreme will to power."[1] Nietzsche characterized this as "Egyptianism." As Nietzsche wrote of philosophers in *Twilight of the Idols*:

> There is . . . their hatred of even the idea of becoming, their Egyptianism. They think they are doing a thing *honour* when they dehistoricise it, *sub specie aeterni* — when they make a mummy of it. All that philosophers have handled for millenia has been conceptual mummies; nothing actual has escaped their hands alive. They kill, they stuff, when they worship, these conceptual idolaters — they become a mortal danger to everything when they worship. Death, change, age, as well as procreation and growth, are for them objections — refutations even. What is, does not *become*; what becomes *is* not . . . Now they all believe, even to the point of despair, in that which is.[2]

Nothing has contributed more to and is more closely associated with this Egyptianism than mathematics. Writing of the illusions associated with claims to truth associated with science, Nietzsche argued that language works to construct concepts. The outcome of this labor is that "the great edifice of concepts displays the rigid regularity of a Roman columbarium[3] and exhales in logic that strength and coolness which is characteristic of mathematics."[4] Pythagoras, embracing the principle of sufficient reason that underpinned Anaximander's cosmology — that there must be a reason why things are the way they are and not otherwise — paved the way for Parmenides to deny the reality of change. The drive to understand the world entirely through mathematics, combined with the principle of sufficient reason, cannot allow for real creativity and the reality of life. All this was clearly evident in Descartes' mathematics and philosophy. The drive to power is now disguised, but it is evident in the dominant quest by physicists to impose on becoming the character of being. The same will to power is implicit.

Many philosophers and mathematicians appreciated where science was leading. It was his appreciation of the nihilistic implications of Newton's mathematical physics that led Kant to argue that the world understood through mathematics is only the world of appearances, not the noumenal world. Kant characterized mathematics as a construction by subjects, developing this conception of mathematics to highlight the paradox of taking the objective world as portrayed by mathematical physicists, the world that appears to have no place for subjects, as the real world. That is, constructivism was developed by Kant to circumscribe and delimit the claims to the validity of science.

Many philosophers and mathematicians have supported this argument, or some variation of it, most notably, neo-Kantians and phenomenologists. Others opposed it. David Hilbert promoted formalism, while Gottlob Frege and Bertrand Russell argued that mathematics should be reduced to logic and defended a purely objectivist semantics that had no place for subjects. Opposing Hilbert and Frege, Luitzen Brouwer, who was influenced by Nietzsche through his teacher, Gerrit Mannoury, defended the role of intuition in mathematics. In doing so, he was not simply defending a particular philosophy of mathematics; he was taking a stand against nihilism, as is evident from his early work, *Life, Art and Mysticism,* published in 1905.[5] Brouwer's intuitionism was another name for constructivism and was really a development of a tradition of thought on mathematics that had begun with Kant's account of and defense of constructivism. As such, intuitionism was also defended by Henri Poincaré and Hermann Weyl, and to some extent by Edmund Husserl, in each case reacting at least in part to the nihilism of mathematical physics.

My claim is that as a process philosopher concerned to overcome the nihilism of scientific materialism, Whitehead had more affinity with these intuitionists than with Frege and Russell, and examining the ideas that influenced him reveals more evidence to justify this view, although these also show the notions of construction and intuition were understood differently. They are more akin to the ideas of C. S. Peirce and are best seen in conjunction with Peirce's philosophy. With this historical background, it should become clearer what characterizations of and developments within mathematics are required to further advance Whitehead's project.

## WHITEHEAD (AND PEIRCE) IN HISTORICAL CONTEXT: WHITEHEAD AND GRASSMANN

Neither Whitehead nor Peirce originally took as their aim to develop philosophies that would overcome the Parmenidean tradition. They were predisposed to do so, however, because each had been exposed to the values that developed in reaction to the nihilism of the mechanistic worldview. This reaction began in Britain and France, but reached its high point in Germany towards the end of the 18th and the beginning of the 19th centuries, with Kant's philosophy being a major source of inspiration for this reaction. Whitehead was exposed to these values through Romantic poetry and the work of the British Idealists. Peirce was exposed to this German influence through the study of Kant himself and through the Concord transcendentalists. However, these values were hardly central to their interests when they began their intellectual careers. There was another source connected to this German philosophical movement, however, a tradition within science and mathematics that, while being inspired by Kant, was much more radical and sought to overcome the whole tradition of Newtonian science. In the case of Whitehead, the crucial figure was the mathematician Hermann Grassmann, although William Hamilton's mathematics and James Clerk Maxwell's physics were also important. The significance of Whitehead's alignment with these thinkers was not fully appreciated by Whitehead or most of his interpreters. What we see in Whitehead is a tension between the influences on him of different traditions of thought, and it is through the working out of this tension in favor of German influences that Whitehead developed as a process philosopher. And in doing so, he exposed with great clarity the flawed assumptions of the dominant traditions of thought.

The tension in Whitehead's philosophy can be found in the Preface to his first published work, *A Treatise on Universal Algebra*. Here he defines mathematics "in its widest signification" as "the development of all types of formal, necessary, deductive reasoning. . . . The reasoning is formal in the sense that the meaning of a proposition forms no part of the investigation. . . . Mathematical reasoning is deductive in the sense that it is based upon definitions which, so far as the validity of the reasoning is concerned (apart from any existential import), need only the test of self-consistency."[6] Although there is a hint of Whitehead's theory of abstraction when he characterizes a "mathematical definition with an existential import" as "the result of an act of pure abstraction,"[7] there is nothing here inconsistent with a logical empiricist's understanding of mathematics, and the construal of mathematics as a system of tautologies. It foreshadows his later effort with Russell to reduce all mathematics to logic, following Frege in this regard. This involved an allegiance to logicism as a distinct philosophy of mathematics defined through its opposition to Hilbert's formalism and to Platonism, but, more fundamentally, to the intuitionism of Brouwer, Poincaré, and Weyl. Frege, following Bernard Bolzano and Hermann Lotze, rejected Kant's constructivism and its implications and was concerned to eliminate any role for mental processes, whether ideas, images, imaginative projections, constructions, or intuitions. As an opponent of intuitionism, Russell denied the significance accorded by Kant to synthesis in perception and thought and rejected Kant's claim that arithmetic is a *synthetic a priori* form of knowledge.

However, Whitehead also characterized *A Treatise on Universal Algebra* as an exhibition of "the algebras both as systems of symbolism, and also as engines for the investigation of the possibilities of thought and reasoning connected with the abstract general idea of space" providing "[a] natural mode of comparison between the algebras . . . by the unity of subject-matters of their interpretation" concerned to provide a "detailed comparison of their symbolic structures."[8] Whitehead acknowledged the source of these ideas in the work of Benjamin Peirce, C. S. Peirce's father. Both Benjamin and Charles Peirce characterized mathematics as "the science that draws necessary conclusions" and regarded mathematics as useful for studying logic, supporting Boole's and de Morgan's conception of symbolic logic as "an algebra of logic" in opposition to Frege's effort, followed by Russell and Whitehead, to reduce mathematics to logic understood as a universal language.

Charles Peirce, together with his father, had made a thorough study of Kant, and later of Friedrich Schelling, who had embraced and further developed Kant's constuctivism. Peirce characterized himself in a letter to William James as a "Schellingian of some stripe."[9] Charles Peirce went on to characterize mathematics through semiosis as "diagrammatic reasoning," treating mathematics as a system of indexical signs the study of which could yield new knowledge. This was a development of Kant's constructivist view of mathematics, not a rejection of it, and gave a central place to intuition associated with observation of diagrams.

After having acknowledged Peirce, Whitehead wrote that "[t]he greatness of my obligation in this volume to Grassmann will be understood by those who have mastered his two *Ausdehnungslehres*. The technical development of the subject is inspired chiefly by his work of 1862, but the underlying ideas follow the work of 1844."[10] Whitehead is unlikely to have been aware of it, but Hermann Grassmann was developing a conception of mathematics advanced by his father, Justus, under the influence of Schleiermacher and Schelling. These philosophers were influenced by Kant, but radicalized and generalized Kant's notion of construction and his ideas on life developed in the *Critique of Judgment*. As Michael Otte argued, "J. Grassmann defines mathematics in the spirit of Schelling, not Kant, as pure constructivity."[11] The Grassmanns' work was part of, and a further development of, the quest to develop a flowing, dynamic mathematics to overcome the Newtonian mechanistic view of the world. Justus Grassmann attempted to develop what he thought of as a "fluid geometry," that is, a "dynamist, morphogenetic mathematics" that would facilitate insight into the emergence and inner synthesis of patterns in nature.[12] It was crucial that this mathematics not be limited to a theory of quantity and be independent of all relations of quantity so that it could go beyond the extrinsic, mechanical behavior of matter and recognize the intrinsic possibilities within nature for structuring and organizing. Hermann Grassmann's work, which he characterized as the "theory of extension," continued this project.

Grassmann presented this work as a survey of a general theory of forms, assuming, as he put it, "only the general concepts of equality and difference, conjunction and separation."[13] He argued that there are two branches in mathematics,

> the continuous form or magnitude [which] separates into the algebraic continuous form or intensive magnitude and the combinatorial continuous form of extensive magnitude. The intensive magnitude is thus that arising through generation of equals, the extensive magnitude or extension that arising through generation of the different.[14]

Grassmann claimed that this second branch was previously unknown, but it is this branch that provides the foundations for all mathematics. As he characterized the aim of his 1844 version of his extension theory in his 1862 reworking of this, it

> extends and intellectualizes the sensual intuitions of geometry into general, logical concepts, and, with regard to abstract generality, is not simply one among other branches of mathematics, such as algebra, combination theory, and function theory, but rather far surpasses them in that all fundamental elements are unified under this branch, which thus as it were forms the keystone of the entire structure of mathematics.[15]

In relation to this, it is significant that William Lawvere, one of the major figures involved in the development of category theory, argued that Grassmann's work was a precursor to category theory.[16]

Extended magnitude was defined by Grassmann as the magnitude created by the generation of difference in which the elements separate and become fixed as separate. This was understood dynamically, as is evident in Grassmann's exposition of the concept of extension theory:

> Continuous becoming analysed into its parts, appears as a continuous production with retention of that which has already become. With the extensive form, that which is newly produced is always defined as different; if, during this process, we no longer always retain what has already become, then we arrive at the concept of *continuous evolution*. We call that which undergoes this evolution the generating element, and the generating element, in any of the states it assumes in its evolution,

an element of the continuous form. Accordingly, the extensive form is the collection of all elements into which the generating element is transformed by continuous evolution.[17]

Grassmann was concerned to apply this mathematics and did so to the study of tides and to electrodynamics. In the tradition of Schellingian thought, the dynamic nature of construction in mathematics was meant to provide insight into the self-construction of nature.

Whitehead's work, insofar as it was influenced by Frege, was antithetical to the Schellingian tradition and supported the development of logical empiricism, a philosophy that cemented in place mainstream reductionist science and rendered any knowledge, apart from how to control the world, as almost unintelligible. However, Whitehead became increasingly dissatisfied with the whole project of *Principia Mathematica,* at least as this project had been understood by Russell. Whitehead saw logic as a means for clarifying mathematical reasoning and exposing defective arguments, but acknowledged that "deductive logic has not the coercive supremacy which is conventionally conceded to it. When applied to concrete instances, it is a tentative procedure, finally to be judged by the self-evidence of its issues."[18] In a late paper, he concluded that "Logic, conceived as an adequate analysis of the advance of thought, is a fake."[19] However, the more fundamental issue was that Whitehead's whole orientation was different from Russell's. While to use Leibnizian terminology, Russell, like Frege, was striving to develop a *Lingua Universalis* — a universal medium whose symbolic structure would reflect directly the structure of the world, Whitehead was concerned to create a *Calculus Ratiocinator,* a method of symbolic calculation which would mirror and refine the processes of human reasoning.[20]

Whitehead's rejection of the project to reduce mathematics to logic and the development of his mature philosophy, and along with it, a different conception of mathematics, was really a development of the early influence of the Schellingian tradition and a creative contribution to constructivist thought. Instead of reducing mathematics to logic, Whitehead argued in "Mathematics of the Good" that "Mathematics is the most powerful technique for the understanding of pattern."[21]

When we say "twice three is six," Whitehead proclaimed in *Modes of Thought*, "we are not saying that these two sides of the equation mean the same thing, but that two threes is a fluent process which become six as a completed pattern." So, for Whitehead, "mathematics is concerned with certain forms of process issuing into forms which are components for further process."[22] Aligned with Schelling, Charles Peirce's characterization of mathematics was also a development within this tradition, and a further advance of it. Combining Whitehead's and Peirce's conceptions of mathematics, we can characterize mathematics as diagrammatical reasoning, studying iconic signs as a way of studying patterns and their transformations, including the patterns of reasoning, and in terms of such patterns and transformations, the relationship between the different branches of mathematics. Diagramatic reasoning is really a form of intuition achieved through the construction and transformation of diagrams. Conceived of in this way, Whitehead's Universal Algebra can itself be seen not only as a form of constructivism giving a central place to intuition and understanding, but as a further precursor to category theory.

## ROBERT ROSEN AND CATEGORY THEORY

It is through category theory that this conception of mathematics can be further developed. Category theory was characterized by one of its proponents as "a powerful language to develop a universal semantics of mathematical structures."[23] The concept of structures is problematic, but Saunders Mac Lane, one of the founders of category theory, characterized mathematics as "not so much about things (objects) as about form (patterns or structures)," virtually equating forms, structures, and patterns. Structures are "lists of operations and their required properties, commonly given as axioms, and often so formulated as to be properties, shared by a number of possibly quite different specific mathematical objects."[24] Category theory enables us to see the universal components of a family of structures of a given kind, how structures of different kinds are interrelated, and to examine the mutability and admissible transformations of precisely defined structures. A category has been defined as "a composite item consisting of a graph and an internal law which associates an arrow of the graph to each path of the graph, called its composite, and which satisfies some axioms given further on."[25] Category theory began with the observation that many properties of mathematical systems can be unified and simplified through a presentation with a diagram of arrows between "objects" (which can be sets, groups, or rings, or can be unspecified), where each arrow represents a function. The most important property of these arrows is that they can be "composed," that is, arranged in a sequence to form a new arrow. The focus is then not on "objects," but on the structure preserving mappings or "morphisms" between these "objects,"[26] These mappings, which reveal the possible transformations of structures, can themselves be studied in this way. If the structures are themselves categories so that the morphisms revealing possible transformations are between categories, these are referred to as "functors," and are represented as arrows between the categories. There can also be a category of functors. The morphisms that transform one functor into another while respecting the internal structure of the categories involved, thereby bringing into focus their mutability, are "natural transformations."

Rosen's conception of mathematics, and its relation to science, is based on his development of category theory as a general theory of modeling. He argued that in fact most mathematics has some referent to something external to the formalism itself, and so is "applied" mathematics, with modeling being the judicious association of a formalism with such external referents.[27] However, category theory makes explicit and clarifies the nature of this modeling relation. Rosen characterized categories as formal systems, morphisms as entailment structures, and natural transformations as explicit modeling of one system by another.[28] He then argued that from material systems we can abstract out "natural systems" which can be modeled in the same way as formal systems are modeled. Modeling natural systems in this way is really hypothesizing via abstractions about their elements and entailment structures to establish congruence between formal systems and these natural systems. This involves carefully delineating observables and linkage relations of the natural systems. There can be no mechanical algorithm for doing this, it is inescapably an art. Once this congruence has been established successfully, we can learn about the modeled system by studying the model of it. This involves using encoding and decoding arrows, along with "dictionaries" to translate back and forth between the two systems, with measurement being a form of encoding, and tracing causal entailments being a form of decoding.

Examining the variety of entailment structures, Rosen argued that modern science, under the influence of Newton, has excluded the kinds of observations, relations, and models with complex forms of entailments that are characteristic of living organisms. Rosen's main concern was to develop mathematics and to reconceive the goal of science to do justice to the reality of life itself. This involved advancing the tradition of natural philosophy inspired by Kant's *Critique of Judgment* and the work of Schelling. Category theory as conceived by Rosen can be interpreted as a major development of the Whiteheadian/Peircian conception of mathematics—as the study through abstraction of possible patterns of connectedness and their transformations utilizing iconic signs or diagrams.[29]

Based on this way of understanding mathematics, Rosen argued that Gödel's theorem is just another foundation crisis for mathematics due to it having taken a fateful wrong turn with Pythagoras. Pythagoras had attempted to reduce geometry to arithmetic, equating effectiveness with an iteration procedure such as counting; that is, computation. It was this assumption that led to Zeno's paradoxes and the crippling of mathematics for millennia. The basic problem is assuming that the simple procedures adequate to simple domains of mathematics, which are adequate for modeling very limited domains of reality, are adequate to more complex domains and can define acceptable procedures. This underpins the quest for formalization, and, over and over again, it has failed. More recent efforts in this direction involved efforts to eliminate semantics from mathematics and to reduce mathematics to syntactical operations without any outside reference. Rosen noted the consequence of this: "once inside such a universe . . . we cannot get out again, because all the original external *referents* have presumably been pulled inside with us. . . . Once inside, we can claim 'objectivity'; we can claim independence from any external *context*, because there *is* no external context anymore."[30]

Most mathematics is not formalizable through axioms as Hilbert called for. For Rosen, what Gödel showed was that the model of arithmetic, developed by Frege, Russell, and Whitehead using set theory and logic, is less rich than arithmetic. Arithmetic is "soft science" relative to the "hard science" of set theory and logic, just as arithmetic is less rich than what is modeled by it, the richness of which is better captured by the "soft" disciplines of the humanities and by the arts.[31] With modeling, this will always be the case. The modeling relation, where something is learned about one system by studying another which is analogous to it, is ubiquitous and characteristic of everyday life as well as of both theoretical and experimental science.[32] It is the failure to appreciate this that has led to the belief that objectivity implies the reduction of biology to chemistry and physics. As Rosen diagnosed source of this problem: "These ideas [that every model of a material process must be formalizable] have become confused with *objectivity* and hence with the very fabric of the scientific enterprise. Objectivity is supposed to mean observer independence, and more generally, context independence. Over the course of time this has come to mean only building from the smaller to the larger, and reducing the larger to the smaller. . . . In any large world, such as the one we inhabit, this kind of identification is in fact a mutilation, and it serves only to estrange most of what is interesting from the realm of science itself."[3]

Once this is realized, we can not only free ourselves from the spell of the Pythagorean/Parmenidean ideal of science, reveal further aspects of its incoherence and free science to acknowledge the reality of life and mind, but develop mathematics more adequate to life. We can also recognize the limits of mathematics and the role and validity of non-mathematical conceptualizations and models that acknowledge some measure of indeterminacy in the present and openness to the future. Rosen showed the real issue and the real problem to be how to develop a mathematical structure in which the logical entailments within the mathematical models adequately reflect the causal entailments in that which is being investigated. In biology, what is being investigated are living beings. Rather than invoke an inadequate surrogate universe, it is necessary to appreciate the full reality

of life itself characterized by final causes and functionality of components.[34] Functional components cannot be fractionated and treated
independently of the organism since they are aspects of and definable
only through the whole organism.

Recognizing a place for final causes, Rosen set out to model anticipatory systems, systems which do not simply respond to their environments but anticipate and respond to what will happen in the future.[35]
That physics at present has no place for the influence of future conditions and final causes indicates, Rosen argued, that it is too specific
and conceptually limited, just as 19th century physics was too specific
and conceptually limited to account for atomic spectra, radioactivity,
and chemical bonding. Just as to explain these required the conceptual
revolutions of relativity theory and quantum theory, so a new conceptual
revolution is required. Rosen questioned the primacy given to closed
systems in science, arguing that open systems are generic, and a closed
system is an extremely degenerate case of an open system. Along with
this, he also questioned the notion of "state" in science and reality,
suggesting that it is a fiction. The conceptual revolution required to
account for life will also require a new mathematics. The mathematics
of Oliver Heaviside and Paul Dirac gave a place to discontinuous signals
(the δ-function), which were initially discounted or denigrated (by John
von Neumann among others) as not genuine mathematics. Nevertheless,
they eventually had to be accepted and the old mathematics relegated
to the status of a special case. Similarly, new mathematics will have to
be developed that will relegate the old mathematics to the status of a
special case.[36] This is what Rosen set out to do.

Modeling anticipatory systems involves modeling systems that produce their own components (in accordance with how Kant and Schelling
understood living organisms). To do this, they require models of themselves (as von Neumann argued). Such systems, Rosen showed, can be
represented through synthetic models in which functional components
are the direct product of the system. In these models the components
are context dependent, and cannot be reduced to fractional parts con-
ceivable independently of the models. Such systems are complex, but not
as mainstream complexity theory understands complexity. This theory,

Rosen claimed, had not freed itself from Newtonian assumptions and dealt only with the complicated. Genuine complexity requires multiple formal descriptions which are not derivable from each other, to capture all their properties. The example Rosen produced to illustrate this was his metabolism, repair, reproduction models (the M-R systems). These models consist of three algebraic maps, one of which represents the efficient cause of metabolism in a cell, another, the efficient cause of repair (that repairs damage to the metabolic processes), and the third represents replication which repairs damage to the repair process.[37] Each of these maps has one of the other two as a member of its co-domain and is itself a member of the co-domain of the remaining map. The maps thus form a loop of mutual containment. As Rosen put it: *"a material system is an organism if, and only if, it is closed to efficient causation.* That is, if *f* is any component of such a system, the question 'why *f*' has an answer within the system, which corresponds to the category of efficient cause of *f*."[38] On the basis of such models it is possible to appreciate the ability of complex systems to incorporate models of themselves in their environments into their behaviour, anticipating future events and correcting their behaviour as new information sheds light on the anticipatory process.[39]

## CREATING A NEW MATHEMATICS

Rosen's work has freed mathematicians from Newtonian assumptions to explore the possibilities opened up by category theory. He has been a source of inspiration for an increasing number of mathematicians and theorists, beginning with his students. A. H. Louie, the most prominent of his students, subsequently published *More Than Life Itself: A Synthetic Continuation in Relational Biology,* and *The Reflection of Life: Functional Entailment and Imminence in Relational Biology.*[40] However, Rosen and his students are not the only mathematicians who have embraced this project of using category theory and, in doing so, have transcended Newtonian assumptions to develop a process relational view of reality. Andrée Ehresmann and Jean-Paul Vanbremeersch in *Memory Evolutive Systems* began by noting that while it is necessary for humans to distinguish objects and their relations, we should not allow ourselves to be dominated by the very limited notion of objects as physical objects located in space; these should include "a musical tone, an odour or an internal feeling. The word *phenomenon* (used by Kant, 1790) or *event* (in the terminology of Whitehead, 1925) would perhaps be more appropriate."[41] An "object" can be a body, property, event, process, conception, perception, or sensation, and it is also necessary to take into account more or less temporary relations between such objects. As Ehresman and Vanbremeersch put it: "Long ago, the Taoists imagined the universe as a dynamic web of relations, whose events constitute the nodes; each action of a living creature modifies its relations with its environment, and the consequences gradually propagate to the whole of the universe."[42] They argued that while Rosen recognized the potential of category theory, he did not fully develop it. They suggest that the role of categories in Rosen's models of metabolism and repair and of organismic systems are often purely descriptive, and do not exploit the deep results of category theory. They claim that in their model, "we make use of fundamental constructions, to give an internal analysis of the structure of the dynamics of the system."[43] They then described their efforts to characterize this in a mathematical model in which "the successive configurations of a system, as defined by its components and the relations among them around a given time, will be represented by *categories*; the changes among configurations by *functors*. The evolution of the system will mostly depend on the interactions between agents at various levels of complexity, acting with different time scales."[44] Memory evolutive systems are multi-scale, multi-agent, and multi-temporal and analyse changes, from an internal, or "endo" perspective, through a net of internal agents acting as co-regulators. Involving a family of categories indexed over time, these are able to model a complexification process internally selected by the net of co-regulators capable of creativity.[45]

The work of Rosen and Ehresmann has stimulated further efforts to develop mathematics adequate to life, notably the non-reductionist biomathematics, or "integral biomathics," exemplified in Plamen Simeonov. Simeonov led the work to produce a major anthology on integral biomathics which he edited along with Leslie Smith and Andrée

Ehresmann,[46] and two special issues of *Progress in Biophysics & Molecular Biology,* the first of which, published in 2013, he edited with Koichiro Matsuno and Robert Root-Bernstein, and the second, published in 2015, with Steven Rosen and Arran Gare.[47]

## CONCLUSION

These developments in mathematics are based on fundamentally different conceptions of what mathematics is and of its role in science than those that led philosophers to deny the reality of change, creativity, and life. Mathematics is no longer assumed to be about what *is*, and only then about transformations, or that it is first and foremost about objects, and only in terms of these, about relations. Also abandoned is the assumption that success in understanding any item in the world is achieved when a largest model can be found from which all other models applicable to it can be deduced, and therefore that the ultimate goal of science is to find the equations modeling the whole universe through which all other features of the universe can be deduced. Furthermore, the Pythagorean assumption that mathematics by itself is capable of modeling every aspect of nature is abandoned. With the new conception of mathematics, we can now view mathematics as playing a major part in comprehending a creative universe rather than explaining away the appearance of creativity. Since living beings are seen to have models of themselves in their environments and can be modeled as such, we can now see more clearly through mathematics, final causes, and activities as transformations, and how mathematical patterns, or forms of definiteness, can ingress in nature. All this requires recognition of the place of synthesis in experience, in mathematical work, in developing models of processes, and in what is modeled through mathematics, and all such synthesis involves constructive intuition. This recognition should free mathematicians to advance further this work of creating a new mathematics.

## NOTES

1    Friedrich Nietzsche, *The Will to Power*, translated by Walter Kaufman and R. J. Hollingdale (New York: Vintage, 1968), §617, 330.

2     Friedrich Nietzsche. *Twilight of the Idols.* Trans. R.J. Hollingdale (Harmondsworth: Penguin Books), 1889/1968), 35.

3     A columbarium is a vault with niches for funeral urns containing the ashes of cremated bodies.

4     Friedrich Nietzsche, *Philosophy and Truth: Selections form Nietzsche's Notebooks of the Early 1870's,* edited and translated by Daniel Breazeale (Trenton, NJ: Humanities Press, 1979), 85.

5     Luitzen Ergbertus Jan Brouwer, "Life, Art and Mysticism," translated by Walter P. Van Stigt, *Notre Dame Journal of Formal Logic* 37.3 (1996): 391–429.

6     UA vi.

7     UA vii.

8     UA v.

9     C.S. Peirce, *Collected Paper* (8 vols), edited by Charles Hartshorne, Paul Weiss, and A.W. Burks (Cambridge, MA.: Harvard University Press, 1931–1966), 6.605.

10    UA x.

11    Michael Otte, "Justus and Hermann Grassmann: Philosophy and Mathematics," in *Hermann Grassmann: From Past to Future: Grassmann's Work in Context,* Hans-Joachim Petsche, 61–70 (Basel: Springer, 2011), 67.

12    Marie-Luise Heuser, "The Significance of *Naturphilosophie* for Justus and Hermann Grassmann," in *Hermann Grassmann: From Past to Future: Grassmann's Work in Context,* ed. Hans-Joachim Petsche, 49–59 (Basel: Springer, 2011), 58.

13    Hermann Grassmann, *A New Branch of Mathematics: The "Ausdehnungslehre" of 1844 and Other Works,* trans. Lloyd C. Kannenberg (Peterborough, NH: Open Court, 1995), 33.

14    Hermann Grassmann, *Extension Theory,* trans. Lloyd C. Kannenberg (American Mathematical Society, 1862/2000), 27.

15    Ibid., xiii.

16    F. William Lawvere, F. William, "Grassmann's Dialectics and Category Theory," in *Hermann Günther Grassmann (1809–1877): Visionary Mathematician, Scientist and Neohumanist Scholar,* ed. Gert Shubring, 255–64 (Dordrecht: Kluwer, 1996), 256.

17    Ibid., 28f.

18    MT 106.

19  ESP 96.

20  These different orientations have been identified by Jaakko Hintikka in *Lingua Universalis vs. Calculus Ratiocinator* (Dordrecht: Kluwer, 1996).

21  ESP 109.

22  MT 92.

23  Andrée C. Ehresmann and Jean-Paul Vanbremeersch, *Memory Evolutive Systems: Hierarchy, Emergence, Cognition* (Amsterdam: Elsevier, 2007), 26.

24  Saunders Mac Lane, "Structures in Mathematics," *Philosophia Mathematica* 4.3 (1996): 174–83, 174.

25  Ehresmann and Vanbremeersch, *Memory Evolutive Systems*, 25f.

26  Robert Rosen, *Life Itself: A Comprehensive Inquiry into the Nature, Origin, and Fabrication of Life* (New York: Columbia University Press, 1991), 143ff.

27  Robert Rosen, *Essays on Life Itself* (New York: Columbia University Press, 2000), 359.

28  Rosen, *Life Itself*, 147.

29  Zalamea, Fernando, *Synthetic Philosophy of Contemporary Mathematics* (New York: Sequence Press, 2012), 219ff.

30  Rosen, *Essays on Life Itself*, 77.

31  Rosen, *Life Itself*, 9f.

32  Robert Rosen, *Anticipatory Systems: Philosophical, Mathematical, and Methodological Foundations* (New York: Springer, 1985/2012), 82.

33  Rosen, *Essays on Life Itself*, 80.

34  Rosen, *Life Itself*, 108ff.

35  See Robert Rosen, *Anticipatory Systems*.

36  Rosen, *Life Itself*, 28ff.

37  Ibid., 248ff.

38  Ibid., 244.

39  Rosen, *Essays on Life Itself*, 199.

40  A. H. Louie, *More Than Life Itself: A Synthetic Continuation in Relational Biology* (Frankfurt: Ontos Verlag, 2009) and *The Reflection of Life: Functional Entailment and Imminence in Relational Biology* (Berlin: Springer, 2013).

41  Ehresmann and Vanbremeersch, *Memory Evolutive Systems*, 21.

42  Ibid., 33.

43   Ibid., 33.

44   Ibid., 21–22.

45   Ibid.

46   Plamen L. Simeonov, Leslie L. Smith, and Andrée C. Ehresmann, *Integral Biomathics: Tracing the Road to Reality* (Berlin: Springer-Verlag, 2012).

47   These two focussed issues are open access issues, cf. *Progress in Biophysics & Molecular Biology* 113.1 (2013): 1–230, and *Progress in Biophysics & Molecular Biology* 119.3 (2015): 205–734.

# 8. Whitehead, Intuition, and Radical Empiricism

GARY HERSTEIN

## INTUITION AND RADICAL EMPIRICISM

It is now 55 years since Eugene Wigner famously articulated his concerns about how it was possible that mathematical models constructed by finite human minds corresponded with reality in such a successful way as to almost seem miraculous. It is not just that mathematical theories were able to effectively model the observed phenomena; often enough, mathematical deductions from the theory would predict an entirely unexpected result that would later be confirmed (consider, for example, the discovery of the positron.) In the end, all Wigner could suggest was that, "The miracle of the appropriateness of the language of mathematics for the formulation of the laws of physics is a wonderful gift which we neither understand nor deserve. We should be grateful for it and hope that it will remain valid in future research."[1] Wigner characterized this last statement as a "more cheerful note" to the pessimism of the rest of his essay.

It seems that very few people took Wigner's concerns seriously. One will occasionally encounter arguments that would suggest evolution and natural selection are sufficient to give an account for how our minds are so fine-tuned to the deep structure of reality. But such claims are abysmally flawed. Precisely what activities did our early hominid ances-tors engage in, such that those activities evolutionarily predisposed their brains to conceive of things like infinite dimensional Hilbert spaces, hyper-complex numbers, homotopic transformations, or Cantorian infinities? Mathematical conception and imagination have so utterly outraced anything that might be supposed to bear even an abstract rela-tionship to the survival skills needed by our evolutionary predecessors that, if anything, Wigner's miracle appears much *more* outrageous, not less, in the light of evolution.

Now, there is a temptation to *overstate* Wigner's "miracle" by assert-ing a much higher success rate to applied mathematical theorizing than is merited by the facts. Nevertheless, some explanation for the connec-tion between human cognition, speculative mathematical reasoning, and natural reality is wanted. As Quine has noted, the fact that we never have more than a finite number of observations at our beck means that we can construct infinitely many mathematical models that will fit those sadly limited collections to any degree of accuracy. How then do those theories that we *do* happen to construct seem to fit the phenomena of the world so well?

Appeals to intuition come to mind, but such appeals do not answer the question. They are, rather like Descartes' appeal to the goodness of God, simply strangling it.[2] The Idealists and the Rationalists appealed to a trans-empirical justificational structure that is somehow "just there," without any additional explanation (but, in the case of Hegel, with an exceptional measure of storytelling, which was not altogether without merit). What *we* want is some account of our connection to reality that at least suggests some empirical sense without starting from a premise that abandons all such sense in the name of a miracle. The account does not have to be perfect, merely credible enough to qual-ify as an account. For this reason, I suggest we shift our attention from mysterious questions about "intuition," to more definite and definitely present issues about *experience*. In particular, I would draw our attention to a system of ideas originally articulated by William James but which are manifestly present in the thought of Alfred North Whitehead, as well. It is in the developments brought by Whitehead to the subject that, I will argue, we find the outlines of an answer for Wigner's "miracle."

In explaining the idea of radical empiricism, one can hardly do better than quote William James on the subject. James tells us that,

> Radical empiricism consists first of a postulate, next of a statement of fact, and finally of a generalized conclusion.
>
> The postulate is that the only things that shall be debatable among philosophers shall be things definable in terms drawn from experience. (Things of an unexperienceable nature may exist *ad libitum*, but they form no part of the material for philosophic debate.)
>
> The statement of fact is that *the relations between things, conjunctive as well as disjunctive, are just as much matters of direct particular experience, neither more nor less so, than the things themselves.*
>
> The generalized conclusion is that therefore *the parts of experience hold together from next to next by relations that are themselves parts of experience.* The directly apprehended universe needs, in short no extraneous trans-empirical connective support, put possesses in its own right a concatenated or continuous structure.[3]

The part that I have emphasized in the above, and to which I emphatically wish to draw attention, is to James' insistence that relations are *real* elements of experience. Experience, in its turn, is real *in its totality*, and not in the cherry-picked elements that are abstracted from experience, such as we find in traditional, and essentially Humean, forms of empiricism.

But the second thing to notice here is that these relations are not merely real "parts of experience;" rather they are *real*. This is part and parcel of the radicalness of James' move in insisting that relations are a real part of experience. Philosophically, empiricism and realism have been at loggerheads for centuries. Yet having never been indoctrinated into any particular philosophical school, Whitehead came to philosophy with his pre-existing science-based commitments to both realism *and* empiricism. *Traditionally*, such commitments were contradictory.

Traditionally, realists said the world — "out there" — was really real, and the only thing on which we ought to predicate our assertions (which, after all, are supposed to be *true*). Empiricists — again, *traditionally* — insisted the only thing we have access to is our own experience. So, on the one hand, claims of truth make no sense unless they are robustly anchored in the world, but no claim about reality can make any sense unless it is robustly anchored in experience. Neither — *traditional* — position could make any headway on this problem.

Among the relations that neither traditional position could account for are the relations of possibility. For both, possibilities are entirely parasitic notions with no reality of their own. They are merely the abstracted flotsam of "genuinely real" actualities. For a radical empiricist, this traditional approach is absurd. We immediately experience the pushes and pulls of possibilities and potentialities. These relations of possibility are the very sinew of experience itself. Possibilities are the stuff that allow the actualities of experience to hang together. Whitehead's critiques of the absurdities of Hume's narrowly traditional empiricism in *Process and Reality*[4] effectively illustrate the fallacies of dismissing relations and relational possibilities. I will have more to say about the contemporary disasters such dismissals have visited upon physical science in the final part of this paper.

It is in his robust approach to the reality of relations that Whitehead's Jamesian approach addresses Wigner's "miracle." Recall again James' three points above. We can rephrase these as, (1) experience is *real*, (2) relations are *in* experience, therefore (3) relations are *real*. A Whiteheadian way of saying this (although Whitehead did not use these exact words) is that nature and reality are all, already, *there*, *in your experience*. Citations justifying this claim are so numerous they almost defy count. For purposes of brevity, let me first draw attention to the discussions of "fact" and "factor" in Whitehead's *The Concept of Nature* and *The Principle of Relativity*. Secondly, I point out that Whitehead continued to endorse his position in these books (indeed, in the entire triptych of works that express his natural philosophy,[5] which includes the above two titles as well as *Enquiry Concerning the Principles of Natural Knowledge*) throughout all of his metaphysical writings.

However, while it is certainly correct to identify Whitehead's position as a very robust form of radical empiricism, it is, perhaps, a bit misleading to use the term "empiricism" with regard to Whitehead. At the time he was writing, the battle in the philosophy of science was between the *traditional* schools of realism and empiricism. And while Whitehead unashamedly referred to James as an "adorable genius," he always emphasized the realist aspects of his philosophy of nature. For this reason, I advocate describing Whitehead's philosophy as "radical realism." It is *radical* because of its Jamesian commitment to the fullness of experience and the "thereness" of relations *in* experience. It is *realism* because that experience embraces the *totality* of *reality.*

So what, then, is our "intuition" into reality, such that mathematical reasoning seems so "miraculous," in Wigner's and many other's eyes? Let me reiterate (in case it was not already clear): I am personally very uncomfortable with the word "intuition." While there is nothing intrinsically wrong with the term, it strikes me that it is used — and used with promiscuous abandon — to mean so many things that it might be better to stop using it at all. Lacking any traces of good sense, I will disregard my own suspicions and continue to use the term. However, like a good Whiteheadian, I will insist on using it in my own idiosyncratic way.

Wigner asked, in essence, "how do we achieve this miraculous bridge across the abyssal chasm between our intuitions (mathematically expressed) and the external world?" The Whiteheadian response is, "We don't, because *there was never an abyss to be crossed in the first place!*" The supposition that there was such an abyss is an artifact of philosophical assumptions that, frankly, do not stand up to any sufficiently careful examination of experience. (I'll have more to say about *this* point in the final section as well.)

In Whitehead's triptych on natural philosophy, he isolates four aspects essential to scientific inquiry. The first is "Fact," which is the totality of unanalyzed nature as it immediately presents itself in its relational fullness in experience. This is the uncompromising holism of Whitehead's realism that makes it so radical. All of Fact is already in experience. The second is "factor." This is the broadest form of selective abstraction from Fact that nevertheless focuses attention upon the salient relational aspects of Fact that are at once *really* there (that is, genuinely in experience) and of substantive importance in some epistemological project or inquiry. The third aspect is "object," which is a functionally stabilized abstraction from factor that retains an algebraically formal "self-identity" across a sufficiently broad temporal spread, such that it can be dealt with in simplified form as a substantive "thing."

The fourth aspect has no direct connection to the relational realities of experience itself. Whitehead calls this a "character." A character—if it is a *real* character—has a definite place within reality. But its derivation and discovery come from the highest levels of cognition, and it is a tertiary form of abstraction that is several steps away from the immediacy of experience. Describing a "character" as a product of "the highest levels of cognition" does not mean that it is necessarily a product of articulated analysis. It might, rather, be the case that it is a product of experience that has been substantially informed by such articulated analysis.

Which brings us back to intuition: As I use the term here, *"intuition" is that deliverance of selected yet extremely hypothetical factors and* characters *that have yet to be tested in explicit analysis, observation, and reasoning.*

In closing this part, let me once again reiterate that Wigner was wrong in supposing there was an abyssal chasm that could only be bridged by a miracle between mathematical intuition and the world. But let me also note, in leading up to the next two sections, that Wigner was also wrong in emphasizing the successes of that mathematical intuition, which he thought could only be found in a miracle. At this stage in our contemporary science, we are in much greater need of attention to our failures. Such are the topics of the next two sections.

## THE MEASUREMENT PROBLEM OF COSMOLOGY

I suggest there is a fundamental problem at the core of the standard model of gravitational cosmology, what is more frequently referred to as "big bang cosmology." I prefer my slightly more awkward phrase because I wish to emphasize that it *is* a "standard"—which is to say, an "orthodoxy. Moreover, it is a *model* (as opposed to the "thing in itself") and about a *gravitational*, rather than a more broadly philosophical, cosmology. All of these distinctions are relevant in the following parts.

Let me begin with a sketch of Einstein's general theory of relativity (GR), which overwhelmingly dominates every aspect and corner of the contemporary standard model of gravitational cosmology. My sketch here is not intended to be sympathetic, only accurate. It is also intended to be *brief,* so I will eschew any attempt at comprehensive thoroughness.

To begin with, GR is what might be called a "monometric" approach. The term "monometric" includes two sub-terms: "mono," meaning "one," and "metric" meaning "measure." This applies to Einstein's GR because the *logically necessary* geometrical *factors* of reality, on the one hand, and the *logically contingent* physical and gravitational *characters* of reality, on the other, are collapsed into a *single measure of reality* which is supposed to capture *all* of the essential relational structures of both, while erasing all of their differences as effectively inessential.

My use of the term "logically" above — in both the phrases "logically necessary" and "logically contingent" — merits some comment. Let me state that I am philosophically "old school" in my use of the term "logic." The notion of "logic" that has dominated the 20th century is one that would reduce the subject to little more than proof theory, a topic which even mathematicians regard as of almost zero value. The approach I advocate here harkens back to Aristotle and finds its contemporary representatives in such persons as Charles Saunders Peirce, John Dewey, and Jaakko Hintikka: *logic is the theory of inquiry*, and formal logic is merely an adjunct to this broader process. The primary use of these latter, formal methods is in indicating what sorts of steps are needed to ask the next round of questions, and what sorts of answers would result in effective falsifications of previously proposed hypotheses. With this in mind, we can enumerate the *logical* (which are not the same as formal mathematical) problems in Einstein's monometric GR. The problems with Einstein's GR are not "mathematical"; rather, they are thoroughly *logical*, in the above-described sense.

The mathematical tools used to represent physically significant relations within GR are known as "tensors." A tensor is a method of compressing a great deal of information into an indexed algebraic form, and then using the interactions amongst the indices of various tensors to derive mathematically rigorous and (ideally!) physically meaningful results. This emphasis in expression is not mere rhetoric: we should keep in mind that Wigner's miracle rotates on just this connection between what the mathematics *says* and what the physical reality *is*.

The step that Einstein took which is so profoundly problematic from a *logical* perspective is his decision to collapse physical and geometrical relations together into a single, metrical tensor. This, of course, is Einstein's famous $g_{\mu\nu}$ tensor, in which the formal characters of space are intrinsically coupled to the contingent physical factors operating at any given point of space, thereby defining the geometrical structure of space local to that point. It is worth noting that Einstein viewed this move, this collapse, as the best part of GR.

The issue here has to do with the logical (as opposed to formal) requirements that are presupposed in any act of measurement.[6] Measurement involves the comparison between a standard of measurement and the thing to be measured. This requires that there be conjugacy relations between the two, such that the comparison is at least *logically* possible. This possibility of direct comparison might only exist in the abstract; yet, if that is so, there will still have to be substantive proxies that can serve as functional adjuncts in the measurement process. And herein lies the problem. With any extensive and/or spatial measurement, the only such proxies that can exist will depend essentially on *the prior establishment of known geometrical relations* between the measurer and the measured.

But such known relations are exactly what Einstein's GR denies us. By collapsing the physical and geometrical relations together, Einstein erased any hope of employing the *logically* necessary relations of geometry as our needed — and, indeed, *only possible* — conjugacy relations for establishing meaningful measurement claims. There are no direct proxies that we can apply to cosmological distances. But since *all* geometrical relations have been folded in with *contingent* physical ones — the very ones we wish to measure, after all — we cannot know the geometrical relations that are *necessary* for grounding our understanding of conjugacy until *after* we already have established the very physical measurements we have no access to. Einstein has created a situation in which we must already know what it is we wish to discover before we have any possibility of actually discovering it.

The above, in compressed form, is the essence of Whitehead's critique of GR. The great tragedy in all of this is that there was never anything necessary in Einstein's decision to *formally* obliterate the *logically* necessary distinctions that would make cosmological measurements at least potentially meaningful. Indeed, quite aside from Whitehead's own alternative — first proposed in 1922 (six years after Einstein's famous GR paper) — there have been a great many alternatives to GR proposed over the years. None of them has ever been examined in a substantive manner by physicists. For example, it was almost 40 years after Whitehead first presented his proposal that any observational evidence could be offered that *might* falsify it. This came in the face of a great deal of potentially falsifying evidence against Einstein that was simply dismissed on the promissory note of further research. This was *forty years* during which time GR was being continuously tweaked, adjusted, and parameterized, so as to *manufacture* the guarantee of its "fit" with observation. I would note in this regard that criticisms of the standard model of gravitational cosmology have been echoed by a number of scientists working in the field.[7]

Yet even as this constant "fine tuning" is allowed to pass without comment by many physicists, upon the discovery of even the slightest deviation from observational evidence, Whitehead's theory was declared utterly and hopelessly refuted. No effort was made, or even allowed, to apply even a part of the "fine tuning" which alone has kept GR viable. The same double standard has been applied to every alternative so far proposed.[8]

The "intuition" — really, "insight" — that Whitehead had and nearly every physicist has ignored since Einstein, is that we have a direct, radical empiricist *experience* of the *uniformity of nature* as part of the immediately given, presented relational realities *in* our experience. Einstein's choice to collapse the geometrical relations, wherein such uniformities are to be found, into the hopelessly contingent physical relations that are supposed to be the ultimate subject matter of physics, was entirely unforced by the empirical facts. Whitehead's argument against this move (for reasons sketched above) was that Einstein's choice was not merely false, but incoherent. To demonstrate this point, Whitehead proposed an alternative theory of his own that exemplified not merely a specific alternative to Einstein and GR, but an entirely different methodological approach. Retaining, contra Einstein, the radical empiricist insight regarding the uniformity of nature, Whitehead separated Einstein's monometrical tensor into two independent measures: Whitehead's $J_{\mu\nu}$ tensor for the contingent physical factors and his $G_{\mu\nu}$ tensor for the necessary geometrical ones.

It is worth pausing to note here that this means the theory Whitehead proposed was unequivocally *bimetric* in nature. Despite this fact, it is physicist Nathan Rosen who is frequently credited with the first bimetric alternative to Einstein, even though Rosen's work came in 1940, eighteen years after Whitehead's. Long after people should have known better, this false narrative continued. Thus Clifford Will in the 1970s caricatured Whitehead's approach as "quasilinear." Will based his work on that of John Synge, and Synge explicitly brags about "throwing away" the philosophical parts of Whitehead's *Principle of Relativity*, where the logic of Whitehead's bimetric approach was so carefully developed. Indeed, when Clifford Will was constructing his "meta-theoretic" system of classifications of theories of gravitation in the 1970s,[9] Nathan Rosen was still alive while Alfred North Whitehead was, alas, still dead. It is difficult to escape the suspicion that Will unconsciously refused to strip the laurels from a living (and famous!) physicist, and hand them over to a dead (and neglected) mathematician and philosopher. Yet it is an irrefutable fact that Whitehead's theory, and *not* Rosen's, is the first *bimetric* alternative to Einstein.

The story of this neglect of Whitehead's work does not end here. Whitehead did not *merely* create a specific alternative to Einstein's GR; he presented an entire *methodological approach*, which he illustrated by suggesting an *entire family* of bimetric alternatives to GR. Only a few pages after he introduces his bimetric tensors, $J_{\mu\nu}$ and $G_{\mu\nu}$, Whitehead brings into play a "mixed" tensor (mixed because it has both super- and sub-scripted indices) $K^\lambda_{\mu\nu}$ which, when applied to his base theory, can modify how physical and geometrical factors can be brought into interaction. Thus, what Whitehead presented is not *an* alternative theory but an *entire family of alternative theories*; the particularities of Whitehead's specific alternative to GR was never anything more than an *exempli gratia* to give concrete form to general family of approaches that Whitehead was genuinely proposing.[10]

To complete this part, let us entertain a question: Can we seriously doubt Whitehead's radical empiricist insight? Can we seriously bring ourselves to believe that nature is *not* uniform in the ways Whitehead emphasized?

The answer to this question must be, it seems, "yes." But that "yes" comes with a severe qualification: "Yes, *provided you disregard experience.*" Because, as Whitehead argued, we experience that uniformity directly, as part of our radical empirical encounter with the "Fact" (Whitehead's term for the totality) of nature. One can disregard this "factor" (aspect) of the Fact of nature, of this manifestly present content of our experience. But doing so comes with a shockingly high price, which more than a few physicists have enthusiastically embraced. They embrace it by prioritizing theory over experience. But their embrasure raises a question: *Quis custodiet ipsos custodes?* "Who watches the watchers themselves?" What standard are we to bring to bear in the evaluation of science, when so many scientists, who have achieved the status of "gatekeepers," declare *ex cathedra* that experience is to kneel before theory? This will be the topic of the next section.

## INTUITIONS, MODELS, AND EXPLANATIONS

The neglect of Whitehead was not total, merely overwhelming. Thus, Sir Arthur Eddington dealt seriously with Whitehead's proposals, which stands out as a considerable achievement given the extent to which Eddington was so singularly responsible for establishing Einstein's GR as an unparalleled triumph within the English-speaking world. Eddington, it is to be recalled, led the expedition to the Indian Ocean to gather photographic evidence of light bending around the sun during a full eclipse. It should also be recalled that the photographic "evidence" that supposedly established the unqualified triumph of GR was based on fewer than a score of actual photographs, more than half of which were thrown away as being unusable. If this looks disturbingly like cherry-picking then, given the history of overwhelmingly dismissive responses to alternatives to GR, perhaps we should be disturbed.

Meanwhile, Whitehead's argument in his 1922 *The Principle of Relativity*, while careful and in many places technical, is not so subtle that its main *philosophical* theses could be missed. That Whitehead's theory was bimetric, that it was a specification of a *family* of theories exemplifying a general approach — these are things that simply could not be missed by anyone who even skimmed the text. And yet, somehow, all of these things and others were, for the most part, trampled over, without so much as a by-your-leave.

All of which begs the questions: How do such things happen in science? *Why* do such things happen? The answers I propose here are conjectural — a thin source of cheer, as the answers themselves are nothing to celebrate. This begins our more detailed tour of what I am calling "model centrism."

Before proceeding, I quote a line from Sir Arthur Conan Doyle that he placed in the mouth of his character Sherlock Holmes on several occasions:

> *It is a capital mistake to theorize before one has data. Insensibly one begins to twist facts to suit theories, instead of theories to suit facts.*

Now, as this statement stands, it is a fabulous oversimplification of matters. One cannot even begin to identify what might qualify as data or facts before one has significant theoretical commitments at hand such as will enable one to identify data *as* data, facts *as* facts. Nevertheless, there is also significant truth in the above maxim that is increasingly lost upon many leading theoreticians in those areas of the physical sciences where data is particularly hard to come by. In this case, that means both gravitational cosmology and high-energy micro-physics. The difficulty inherent in these disciplines, of gathering independent lines of evidence and then interpreting that evidence, encourages "model centrism": the conviction that implicitly (and, in some cases, explicitly) holds that it is sufficient for science to have a model that "accounts for" the data, without making any effort to explain or appreciate how the model achieves that "account."

This is a deeply problematic attitude; indeed, it is the very paradigm of pseudo- and even anti-scientific thinking. This is because the sorts of compromises that can be built into a "model" become so egregious as to render any and all data effectively moot, because any real possibility of *testing* the model has been written out of the situation through the promiscuous addition of freely adjustable parameters that guarantee in advance that a "fit" with observation can be manufactured regardless of the actual content of those observations. Model centrists are persons who, while at least nominally scientists, in actuality disregard this catastrophic problem with their models in favor of becoming triumphalists and cheerleaders. The "intuitions" about the model, if you will, take complete priority over the empirical facts.

Some leading model centrists include Stephen Hawking, Brian Greene, and Sean Carroll. These scientists have variously disdained observation over theory (Hawking, Greene),[11] called for an end to (or at least a radical weakening of) the falsifiability and testability criteria for scientific theories (Carroll),[12] and denounced philosophical critique as having no place in the modern world (Hawking). On this last point, Stephen Hawking stands out as the most painful example, not so much for his explicit denunciation of philosophy, but for his blatant engagement in philosophical speculation that he evidently believes is an example of scientific exposition. However, while Hawking's (and others) efforts in this regard are, on the one hand, truly abominable philosophy; on the other hand, it is not really "bad" science because it is not any kind of science whatever. Nothing they claim is testable, and whatever observations come along, it is always possible to adjust the parameters of the model so that said observations go away.

I've already mentioned that not all scientists are happy with this state of affairs.[13] In addition, physicists such as Lee Smolin and Peter Woit have emphasized the empirical and scientific vacuity of string theory as well as contemporary cosmology. Here again, we have a *model* that is accepted because it is mathematically—which is *not* to say *logically*—interesting, but for which no manner of test is possible. Once again, the attitude amongst many of the contemporary gatekeepers of physics seems to be, "why trouble ourselves with data, when we have a clever theory?" What sorts of intuitions are guiding these scientists? Well, clearly they are not empirical ones, radical or otherwise.

While much is to be said in favor of a well-educated "intuition"—which is to say, an attentiveness to the full relational contents of our radically empirical experiences—there is also much to be said against an "intuition" that has been misdirected—if not downright mutilated—by a priori assumptions that valorize theory over experience, whatever the cost. Yet how are we to tell when our intuitions are of the one type rather than the other? The answer, in some ways, is so obvious it is almost painful: Just ask yourself, "What, ultimately, is the basis of our intuitions, theory or experience?"

This choice becomes less of a problem when we embrace experience in its *radically relational fullness*. No more disregarding experience in its completeness, no more crippling the interpretation of experience with 17$^{th}$ and 18$^{th}$ century psychologies. In this regard, let us remind ourselves of the conclusion from the first section of this paper: the problem with mathematics and reality is *not* the one that Wigner identified, of how we "bridge" the supposedly impossible chasm between thought and reality. The problem, rather, is how do we *select* those characters and relations that are already *in experience* which are genuinely representative of nature? Most of what has followed that conclusion has been devoted to saying what we *must not* do and arguing that the blind alley of model centrism does not serve any meaningful concept of science.

Einstein's "intuitions" of the world had a pronounced aesthetic component that idealized mathematical elegance over robust empirical content. In addition, Einstein understood a great deal about marketing, and as a result his particular "intuition" of "how the world is" came to be rigidly embedded in our cosmology.[14] There is nothing necessary about this result, about this peculiarly aesthetic vision of Einstein's and its standing in science—nothing either empirically, nor scientifically, certainly not ontologically, and most certainly not logically necessary.

Yet it has come to impose itself not only on how we actually do envision the real, but upon how we are and even might be permitted to envision the world. The standard model of gravitational cosmology is not merely the model we currently have, it is the "measure" of what sorts of models we will allow to be seriously considered.

Toward the end of *Modes of Thought,* Whitehead remarked that "Physical science has reduced nature to activity, and has discovered abstract formulae which are illustrated in these activities of Nature. But the fundamental question remains, How do we add content to the notion of bare activity? This question can only be answered by fusing life with Nature."[15] Whitehead's radical empiricism is the recognition of that fusion at the cognitive level, where the lively relational fullness of experience becomes the solid basis for our intuitions about the world.

Intuition, it should be noted, *must* be cultivated in order to develop the sensitivities needed to zero in on those characters of the world that are to be found in our radically empirical experiences of nature, and which are salient to our various inquiries. But while we do not want our intuitions to be a wild patch of ground, gone entirely to weed, neither do we want it to be some insanely "bonsai'd" garden that has been artificially twisted into some egregiously rococo pretzel. Quoting Emerson here, "How wearisome the grammarian, the phrenologist, the political or religious fanatic, or indeed any possessed mortal whose balance is lost by the exaggeration of a single topic. It is incipient insanity. Every thought is a prison also."[16] So when we thoroughly canalize our intuitions in a manner that valorizes mathematical models over living empirical data, we are stepping in a trap that we ourselves have built for our minds.

And herein we find the second response to Wigner, a far less happy one than even leaving things at just and only an inexplicable miracle. The "fabulous success" of mathematics in the natural sciences is, in many cases, an illusion raised to the status of dogma by the model centrism that is so rife in contemporary physics. Our "successes" with mathematics have been frequently guaranteed at the outset by our aggressively parameterized models that manufacture their "fit" with reality rather than discover it. There is no finite bound to the mathematical models we can generate in order to ensure the "fit" of our embarrassingly finite observations.

I will conclude, then, by observing that Whitehead's intuitions about space and time, as well as science in general, are ones of modest humility. In his intuitions, we do not approach the world with a collection of assumptions that undermine the very possibility of measurement, nor do we idolize mathematical cleverness for nothing more than its own sake. It is our theories that must fit the data, not the other way around. And we must not hypothesize beyond what we can test and then treat those hypotheses as given facts. (Whitehead famously quipped in this regard, "seek simplicity, and distrust it."[17]) While such rationally constrained science might lack the Roman spectacles of M-theories, multiple and multiplying universes, and twisty-turny space-time, it has the virtue of leaving discussions of angels dancing on the heads of pins aside, so that we might focus on genuinely achievable facts.

## NOTES

1    Eugene Wigner, "The Unreasonable Effectiveness Of Mathematics in the Natural Sciences," *Communications in Pure and Applied Mathematics* 13.1 (1960).

2    I am shamelessly stealing this analogy from Ernst Cassirer.

3    William James, "Preface," in *The Meaning of Truth*, multiple editions, emphasis added.

4    Whitehead's criticisms are far too numerous and well known to make citing any one example necessary.

5    For reasons I'll not go into here, I also advocate speaking of Whitehead's approach as "natural philosophy," rather than "philosophy of science." I'll merely suggest that Whitehead is more "old school" than the rather contemporary notion of "philosophy of science."

6    My discussion here will be limited to spatial and (more generally) extensive measurement. There are subtleties in other types of measurement that need not concern us. See Krantz, Suppes, et al, *Foundations of Measurement*, volumes 1–3 (Mineola: Dover Books on Mathematics, 2006).

7    See, for example, Michael Disney, "Modern Cosmology: Science or Folktale?" *American Scientist* 95.5, (2007): 383. Also, Chapter 6, "Doubts About Big Bang Cosmology," in *Aspects of Today's Cosmology*, edited by Antonio Alfonso-Faus, Intech, DOI: 10.5772/1838. See also Paul Steinhardt,

"Theories of Anything," *Edge 2014: What Scientific Idea Is Ready for Retirement?* https://edge.org/response-detail/25405 verified August 31, 2015.

8    For a number of alternatives, use the following URL: search.arxiv.org:8081/ ?query=bimetric&in=grp_physics.

9    Clifford Will's "Parameterized Post-Newtonian Framework," in *Theory and Experiment in Gravitational Physics*, Revised Edition (Cambridge: Cambridge University Press; 1993).

10    See Gary Herstein,*Whitehead and the Measurement Problem of Cosmology*, (Frankfurt: Ontos-Verlag [now De Gruyter], 2006).

11    Stephen Hawking's most recent bromide is *The Grand Design*, (New York: Bantam Books, 2012), where he advocates for "model-dependent realism" as an "escape clause" (my term) from the empirical vacuity of contemporary theories. Brian Greene, *The Fabric of Reality*, (New York: Random House, 2007) is equally dismissive of empirical content in favor of clever theories.

12    Sean Carroll, "Falsifiability," found in *Edge 2014: What Scientific Idea Is Ready for Retirement?* https://edge.org/response-detail/25322 verified September 12, 2015.

13    See note 7 on Michael Disney and Paul Steinhardt.

14    See, for example, Jimena Canales, *The Physicist and the Philosopher: Einstein, Bergson, and the Debate That Changed Our Understanding of Time* (Princeton: Princeton Universaty Press, 2015) for a discussion of the social and historical factors behind the acceptance of GR.

15    MT 228–29.

16    Emerson, "Intellect," in *Essays: First Series*, multiple editions.

17    CN 163.

# 9. *What Does a Particle Know?*
# *Information and Entanglement*

ROBERT J. VALENZA

M y two presentations at the 10th International Whitehead Conference addressed materials that had been published in two prior papers. The first, "Possibility, Actuality and Freewill," (2008) was a reduction of the freewill theorem of Conway and Kochen (2006) to something mathematically and conceptually homomorphic to their much more technical combinatoric analysis. The second, "What Is It Like to Be a Photon?" (2015) was a later attempt to resolve an apparent inconsistency of special relativity with process metaphysics. I will say just a little more about the substance of these papers below, but for now I wish only to observe that each concerns fundamentally the concept of information. Thus, for example, the freewill theorem speaks of the evolution of the state of an elementary particle as being undetermined by *all of the information available to it in the universe*. As so often happens in philosophy, a simple word that we use effortlessly can sometimes be called onto the carpet for deep questioning. In this case, the very idea of information, which is central to both large- and small-scale physics — that is to say, both to relativity and to quantum theory — is perhaps used somewhat too nonchalantly. The purpose of this short essay then is to elucidate the concept, especially in light of the two papers I presented, and hence the title. I can at best present an introduction to the issues, for in searching the literature on information, one will find a vast wealth of material in mathematics or communications theory, but only subject to a prior characterization of information that is philosophically unsatisfying. When we descend into the more metaphysical and speculative realms, far less is available. Nonetheless, we do have two touchstones: we can check what we are proposing against the communications-theoretic interface and against the intuitive use of the term as adopted by the community of physicists. With this in mind, we shall try out a paradigm and see how it fares.

## A TRIPARTITE CHARACTERIZATION

My basic theme is that there are three essential and more fundamental aspects to the concept of information, and these are strongly connected. The three aspects correspond to the notions of *physical substrate*, *abstract structure*, and *semantic efficacy*; these, and their relations, are illustrated in Figure 1.[1] Each node and connector on the diagram requires considerable explication, but let me make some immediate general remarks before any such effort. First, the three nodes of the diagram do not represent any sort of hierarchy: they represent ideas at the same level, as reflected by the imposition of a common textural background and the circularity of the arrows in the composite. That these three are given, in a sense, on equal footing is tightly bound to a rich and controversial set of metaphysical implications, as we shall see below. Indeed, we see here something of the reciprocity of which Robert Rosen speaks in discussing the inertial (that is, passive) and gravitational (that is, active) aspects of assemblies such as genes. (2000, 13; Kindle location 372) Second—and very much related to this last point—the diagram is not intended to project closure among these three nodes. There must be other figures in play on the canvas, but because I am specifically making the effort to elucidate the idea of information to distinguish it from other kinds of things in the world, I emphasize only these nodes, just as a physiologist interested in the respiratory system might include the lungs but not the kidneys in a diagram. She or he would thereby not be denying the existence of the kidneys or their ultimate relationship to the lungs, but merely acknowledging a kind of conceptual metric. Finally, the terms attached to the arrows in Figure 1 may be read informally with some technical points to follow. Moreover, there are some tempting and even obvious connections with process metaphysics, but I shall not, in this limited space, do much more than suggest these connections, making no attempt to fold them into any comprehensive theory of concrescence, although I believe this would not be difficult.

Figure 1. A tripartite characterization of information.

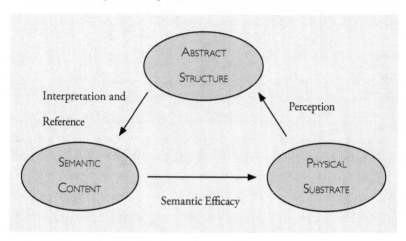

## THE PHYSICAL SUBSTRATE

Our notion of the physical is so vivid that it seems to require little or no elucidation. Nonetheless, since I would like to give it a functional characterization for this discourse, a few words are in order before discussing the relationship of physicality to information.

Why is the word object embedded in the word objective? We might begin by saying that the objects of the world are to a great extent what we can speak of in the third person—with the expectation that others will also speak of them, or at least nod comprehendingly—and rests reciprocally on a kind of perspective invariance, a concept for which I would give general credit to Thomas Nagel. A variation on this would be the tightly associated notion of context independence, which is directly attributable to Rosen.[2,3] My point here is that both of these characterizations shift physicality away from ontology and somewhat toward epistemology: it is not that the tree is there whether anyone is looking; it is that the percipients agree on the experience and subsequent report of a tree. A general benefit of this emphasis on intersubjective coherence is that by itself it vitiates some of the poison of dualism. (Of course, there is no such problem within the discourse of process metaphysics.) The particular benefits for our characterization of information lie in connection with the right and middle nodes of Figure 1.

In analyzing the notion of data as it pertains to information, Luciano Floridi makes this, to my ear, rather profound assertion: "[A] datum is ultimately reducible to a lack of uniformity." (2010, 23, Kindle loc. 453) Since Floridi is speaking of the physical world here, for brevity, and in consonance with an essential element of process metaphysics, I shall simply speak of data as physical contrasts. There are other constraints, as we shall see, but for the moment, the idea that these contrasts should be definite—or definite enough—amounts to a requirement of perspective invariance in accordance with our characterization of physicality. In its barest form, this requirement of contrasts reduces to a binary state: dash or dot, black or white, stimulus or no stimulus. Implicitly, this invokes subjectivity in the sense of there being a percipient, but, as just noted, this implicates nothing new ontologically. More importantly, whatever else we might say about data as entailed by the notion of information, this grounding in physical contrasts does make the required connection with the often tacit axiom of information theory that its subject matter admits a model consisting of binary strings.

To summarize, the right-hand node of our figure and the arrow pointing away from it begin our characterization of information as a matter of the perception of definite contrasts hosted in a physical substrate. To combine Whitehead with Rosen, one might say that the inertial aspect of information rests with actual entities. This does leave us with two layers of ambiguity, neither fatal: Do we want to say that the physical substrate itself contains information if it is not being processed by perception? Do the audio disks lying on my shelf indeed contain information? More particularly, does it count as information if the disk is being played but no one is listening? With respect to the first question, we might speak of *potential information,* which is to assert the possibility that the physical substrate may take part in the complete set of relationships indicated in the figure. With regard to the second, for reasons that I will address below, I would say, yes, potential information counts as information whenever it is being processed.

Finally, note that it is not my intention that the right-hand node of Figure 1 represent a single physical system, such as an audio disk. It represents the physical world, with the word "substrate" emphasizing that parts of the world serve as a kind of platform for information. This distinction will be important later in understanding the horizontal arrow in our picture.

## ABSTRACT STRUCTURE

While we have begun our characterization of information at the level of the physical, another immediate aspect is the attribute of detachment. Consider this assertion, again from Floridi:

> Despite some important exceptions (e.g. vases and metal tools in ancient civilizations, engravings and then books after Gutenberg), it was the Industrial Revolution that really marked the passage from a world of unique objects to a world of types of objects, all perfectly reproducible as identical to each other, therefore indiscernible, and hence dispensable because replaceable without any loss in the scope of interactions that they allow. (13; Kindle loc. 357)

While he is not speaking directly about information here, Floridi is certainly nibbling at the idea of the abstraction of form from a particular instance of form. This in itself might go back to discussions of Plato and Aristotle, but consider Floridi's overall context: speaking directly about information now, the point is that in processing the data from the raw substrate, information admits a kind of detachment from that substrate. Lyre (2002, Chapter 1) might identify this element with a syntactical requirement. I do not deny that in the least, but Whitehead gives us something much more directly apropos.

I characterized Whitehead's idea of *conceptual reversion* in a previous paper (Valenza 2002):

In process thought, conceptual reversion denotes a specific feature of concrescence that one might summarize as follows. In the transition from the conformal to the conceptual phase of concrescence, physical feelings engender conceptual feelings, and this association may be thought of as a kind of fusion. The eternal objects that serve as the data of these conceptual feelings admit in turn a flexible and creative association with other eternal objects, as mediated by the primordial nature of God. These derivative eternal objects then admit a second fusion with the physical feelings, giving rise to a *reverted* conceptual feeling. By its very nature, this secondary fusion may lift these physical feelings away from their origins; or, to use Sherburne's felicitous phrase (1961, 168), prevent them from "sink[ing] back into immanence."

My point here is that the possibility of conceptual reversion, or something very much like it, is necessary in the perception of the contrasts associated with informative data: we must be able to lift out its abstract structure. One sees here an enormously strong affinity with the notions of syntax and syntactical entailment in the linguistic sense of these ideas, but conceptual reversion is a larger concept, and, unlike syntax, one not specifiable in advance.

Let me refine the abstraction required by the top node of Figure 1 by introducing a special term that came into broad use with Einstein, one that has been identified as implicit in process metaphysics (see Stolz 1995) and is rather more explicit in the work of Rosen (1991) in his direct appeal to category theory: that term is *covariance*. Covariance is an aspect of abstraction or pattern that is a kind of generalized symmetry. Specifically, for Einstein it was the idea that the formalisms that describe the universe should be independent of the frame of the observer. More generally, it is a statement that there is no privileged node of subjectivity. With regard to my use of abstract structure as an element of information, I am asserting that the pattern is covariant, and, in the particular sense discussed above, this assimilates to an aspect of objectivity.[4]

Thus, at this point we have asserted that information consists of a perceptible, covariant pattern of contrasts that are physically realizable.

Moreover, in the context of Whiteheadian metaphysics, a key component in this process of abstraction is conceptual reversion.

## SEMANTIC CONTENT

The left node and the arrows in and out of it remain to be explained. More so than previously, the arrows and the node must be understood together.

Clearly, the figure indicates that semantic content is the result of interpretation and reference. For our purposes, the notion of reference suffices, and interpretation is simply an auxiliary notion that helps informally to elucidate the technical sense in which reference is applied here. Both the word "reference" and the context bring to mind Whitehead's notion of *symbolic reference*, by which, in its less abstract metaphysical sense, Whitehead denotes, "[t]he organic functioning whereby there is transition from the symbol to the meaning" (S 7). I have argued (2009) that this reduces to the association of a pair of experiences, and this is very much the spirit of what I intend by the left arrow and its target in Figure 1. Thus, information connotes some recognizable abstract pattern, but this pattern in turn must be subject to an association with a component of an actual occasion. Indeed, we may regard this as a kind of syntactic requirement associated with the perception of a pattern of contrasts as information. However, there is more: this association with an actual occasion must itself be subject to another kind of transformation that we speak of as *semantic efficacy*. Since I wish to distinguish this from ordinary causal efficacy in the Aristotelian sense, let me exemplify what we have said so far with an effective example borrowed from Searle (1998, Chapter 5).

Consider an ordinary one-dollar bill as a bearer of information. Unremarkably, the lifting of its abstract structure is a matter of manifold contrasts: at perhaps its crudest, this amounts to separating the object from its environment *as a single object*, and at a much more refined level there is a complex matter of pattern recognition and being drawn to the Gestalten that represent letters and numerals in the English language. Clearly a great deal of this "processing" might be a matter of physical description, but, except to the most devout materialist or physicalist, this object possesses, via reference, a semantic efficacy that is beyond our power to attribute to its physical attributes.[5] In that specific sense, the *information* carried by a dollar bill (as opposed to its raw physical description) is a matter of effective semantic content, in this case achieved via a social construction. This is exactly what we intend by semantic efficacy: a species of efficacy that is not described by the language of physics.

## IN-FORMATION

This last matter as represented by the bottom arrow of the figure is so critical to my point that I should like to give it still another name and characterization: *in-formation*. The ordinary word remains as a substring, and I wish in that to retain everything we have said so far that distinguishes information from the mere physical disposition of some part of the world; the hyphen emphasizes that part of the root concept that operates in the ordinary physical world and yet clearly seems to bypass mere physical efficacy. Social constructions and symbolic transactions are thus very much comprised of information insofar as they are *in-formative* in this specific sense of semantic efficacy. And while I have not attempted to embed this brief theoretical sketch into the full technical panoply of process metaphysics, it most evidently accords with it. Stated briefly, I would say that the idea of information as in-formation is implicated in the notion of perception in the mode of presentational immediacy (with its sharp contrasts) and in the ultimate ingression of necessarily prior actual entities in the becoming that is concrescence.

## THE ENTANGLED PHOTON OF THE FREEWILL THEOREM

While admittedly nothing dispositive is achieved here, I did set two goals at the outset: the first was to tie whatever characterization of information was developed back to the sense in which it is used in communications theory, which is to say, something isomorphic to a binary string; the second was to test the characterization for consistency with how the notion of information is used by physics, in particular in connection with the theorem on freewill by Conway and Kochen (2006). I will now make two points germane to the latter.

My first point is that in consideration of Figure 1, a physicist might want to assert axiomatically that the arrows representing perception and semantic efficacy cannot be dissociated from the transfer of matter or energy, and therefore whatever effects they implicate are limited by the speed of light. Even if he or she is willing to accept the idea that more is inherent in the notion of information than direct physical efficacy, and moreover accepts that the notions of abstraction and semantics must be admitted to complete the description, I still find something metaphysically unsatisfying about this position: it has built into it a complex physical contingency, namely the speed of light in a vacuum. Moreover, it seems to me that if we allow matters to rest here, it becomes tempting to revert to mere physicalism mediated by this limitation.

My second point makes a more compelling case for the characterization of information above, but in doing so, I need to introduce a couple of brief technical statements from the proof of the theorem by Conway and Kochen. As explained by them (and explained in simpler terms in my 2008 exposition), the crux of their proof that if humans have freewill, then so do, say, photons, is this: their analysis shows that a decision made by an observer at some point A and the subsequent measurement of some attribute of a nearby particle is sufficient to conclude that not all of the information available in the universe at some distant point B is enough to determine the disposition of an entangled particle with respect to a similar measurement. The key is that the information on the decision made by the observer at A cannot possibly reach points near B in time if the distance is great enough. Thus insofar as that decision influences the measurement at A and in fact entails the measurement at B, we can speak of the particle at B as having the same degree of freewill as the observer at A.[6]

With this background, what I want to emphasize here is that once this theorem is in place, there is an obvious symmetry among the events at A and those at B. We could equally say that not all of the information available at A is enough to determine its disposition with respect to the measurement in question, but, nonetheless, the measurements at A and B are coupled. My main point then is that, quite independently of any top limiting speed in the physical universe, *Figure 1 has a directionality*.

The processing of information is cyclic, and whether we think of the implied temporality as objective or subjective, this cycle is not symmetric with respect to time. Thus no matter the joint-entailment of the measurements at A and B, there can be no sense in which the events at either point in-form those at the other. Metaphysically, I find this to be more satisfying in asserting that whatever connects the events at A and B does not count as information.

## THE SELF-REFLECTIONS OF A PHOTON

Finally, I mentioned two papers in the introduction, the other not concerned with freewill but with the reconciliation of the photon as an integrated society. The problem in a nutshell is this: Since physical time does not pass for a free (non-interacting, real) photon, how does it experience its past in progressing from one instance of an actual entity to the next? It would seem as if the photon must survive without the information needed to support its continuity as a society. After some analysis, the proposal set forth in that paper in part depended on the presumption that perception in the mode of causal efficacy was not intrinsically spatiotemporal, as is perception in the mode of presentational immediacy. Accordingly, let me make a brief comment on how this meshes with the theory of information developed here. The point that must suffice is the distinction Whitehead makes between these modes of perception: the latter is one of sharp contrasts, while the former is a matter of vague feelings. The upshot is that causal efficacy does not support the requirement that information, as it is perceived in the physical substrate of the world, admits the unambiguous contrasts on which the abstraction of structure depends. Hence, here we need no subtle symmetry argument to see how our notion of information — or lack thereof — plays in the analysis of the society of an uninformed photon. Something rather simpler suffices, and this comes with the physical atemporality of the photon and Whitehead's own characterization of perception in the mode of causal efficacy.

## NOTES

1    Although the characterization that I am suggesting here was developed independently, there is a striking resemblance to that given by Holger

Lyre (2002, in German). He states early on: "The complete concept of information emerges in the dimensions of syntax, semantics, and pragmatics." (17; translation mine) Later, following Weizsäcker (1985), he says, "Information is only what it produces." (19) Thus the role pragmatics plays in his characterization is similar to what I have identified as the physical substrate together with semantic efficacy. Moreover, syntax as pattern also falls under my rubric of abstract structure, although I intend something more general here.

2    Following his introduction to inertial and gravitational as metaphors for the passive and active roles of certain objects, Rosen concludes with respect to the usual scientific meaning of these concepts, "Roughly, these serve to couple states or phases (i.e., whatever is behaving) to forces. In mechanics, these parameters are independent of both phases and forces, independent of the behaviors they modulate. Indeed, there is nothing in the universe that can change them or touch them in any way. Stated another way, these parameters are the quintessence of objectivity, independent of any context whatever." (ibid.)

3    Are perspective invariance and context independence equivalent markers of objectivity? It would seem to me that the latter entails the former insofar as perspective is an aspect of context. Whether the converse is likewise true depends on a more delicate explication of the notion of context than we are prepared to give. Moreover, if context independence properly subsumes perspective independence, we might still ask whether we need something this strong to characterize physicality. At this point, I would conjecture that context independence is both broader than perspective invariance and too broad for physicality.

4    Note that here again the assimilation of objectivity to perspective invariance is most helpful in resolving a fundamental paradox in mathematics: How can idealizations such as numbers, circles, and triangles seem so objective when, in the physical world, there are no such things? Why are these among the easiest words to learn in a foreign language? Because they are context invariant, and, in particular, perspective invariant. This speaks directly in support of the conjecture raised in the previous footnote.

5    I am certainly aware that our devout physicalist might want to maintain that the physical perceptions involved here simply propagate through the neural machinery and motor systems of the human organism, and therefore we need not leave the world of third-person physical descriptions to describe the effects of this complex artifact. He or she might indeed want to maintain this, but we still have no complete description—just dogma and scientific orthodoxy—of how it all occurs. With respect to our current discussion, we are treading lightly on the problem of dualism

here, but in discussing abstraction in nonphysical terms, we have already acknowledged something beyond the physicist's normative assumption that the world is nothing more than matter-energy in motion. In fact, the title of the chapter in which Searle raises this example (for a somewhat different purpose) is "The Structure of the Social Universe: How the Mind Creates an Objective Social Reality."

6    I emphasize that this is a very particular notion of freewill that has nothing to do with the ordinary modality of intention.

## REFERENCES

Conway, John and Simon Kochen. "The Free Will Theorem." *Foundations of Physics* 36.10 (2006).

Floridi, Luciano. *Information: A Very Short Introduction.* Oxford: Oxford University Press, 2010. Kindle Edition, 2010.

Lyre, Holger. *Informationstheorie: Eine philosophisch-naturwissenschaftliche Einführung.* Munich: Wilhelm Fink Verlag, 2002.

Rosen, Robert. *Life Itself.* New York: Columbia University Press, 1991.

Searle, John R. *Mind, Language and Society: Philosophy in the Real World.* New York: Basic Books, 1998.

Sherburne, Donald W. *A Whiteheadian Aesthetic.* New Haven: Yale University Press, 1961.

Stolz, Joachim. *Whitehead und Einstein.* Frankfurt am Main: Peter Lang GmbH, 1995.

Valenza, Robert J. "Aesthetic Priority in Science and Religion." *Process Studies* 31.1 (2002).

Valenza, Robert J. "Possibility, Actuality and Freewill." In a special issue of *World Futures Journal*, edited by Franz Riffert and Timothy Eastman (2008).

Valenza, Robert J. "Meaning and Process." In *Chromatikon V, Yearbook of Philosophy in Process.* Ed. Michel Weber and Ronny Desmet. Louvain-la-Neuve: Presses Universitaires de Louvain, 2009.

Weizsäcker, C. F. von. *Die Einheit der Natur.* Munich: Hansen, 1971.

## *10.*   *A Neurobiological Basis of Intuition*

JESSE BETTINGER

As we continue to assess the degrees to which Whitehead's logic can be interlaced into models underwriting physical law in nature (e.g., Eastman 2008; Bettinger 2015), so too should we be aware of the extent to which this theory (1929) of actual entities (AEs) precipitates the notion of a biological counterpart. The aim of this essay is to introduce that counterpart and then correlate it to key concepts entailed in Whitehead's theory of perception (1927; 1929). Whitehead developed his theory of AEs in the context of both the emerging field theories of physics and the science of physiological psychology, calling it "a cell theory of actuality" (PR 242); and yet, while promising preliminary work (e.g., Riffert and Weber 2003, Northoff 2009, Weber and Weeks 2010) has certainly set the stage, room remains for a neurobiological platform of Whitehead's AEs to be enunciated with even more rigor. Following Georg Northoff, a two-part question can be posed: 1) can neuroscience further inform an understanding of Whitehead; and 2) can we acquire insights into neuroscience from a Whiteheadian perspective? This essay will identify affirmative answers for both and, in the process will provide a neurobiological framework that can be readily applied back into a model of renormalization physics (à la Verlinde

2011; Bettinger 2015), thereby tying the knot between intuition, math, and physics.

The first part of this essay introduces the faculty of "interoception" as the ability to sense the internal conditions of one's body. This sense is entailed in a summary description of the neurobiological architecture and dynamics of the brain-gut information pathway (the "neurovisceral axis"). Entailed in this axis are the evolutionarily rare "von Economo neurons" (VENs), named after the Austrian psychiatrist and neurologist who emphasized their uniqueness in 1925. These specialized neurons facilitate the rapid information flow of signals from the gut to the brain and are considered to facilitate a neurobiological basis of intuition (Allman et al. 2007) or literally, of gut feelings.

The second part of this essay correlates key features of the neurobiological description of part one with major concepts of Whitehead's theory of AEs in the capacity of his theory of perception. These neurobiological dynamics are correlated with the dynamics of Whitehead's *prehension, concrescence,* and *satisfaction,* as well as with his theory of perception in the three modes of *causal efficacy, presentational immediacy,* and *symbolic reference.* Additionally, Ralph Pred's 2009 book, *Onflow,* provides tremendous support by identifying what he calls a "concrescual on-flow" of information, or signal processing. This line of reasoning prompts a review of bottom-up signal processing from a mere additive accumulation to a selective synthesis of values: "a concrescence of prehensions" (PR 35). By the end of the essay, it should be apparent that there is a symbiotic connection between Whitehead's process philosophy and modern advances in the neuroscience of interoception and the neurovisceral axis.

## INTEROCEPTION AND THE NEUROVISCERAL (NV) AXIS

We can all perceive feelings from our body related to its internal and external state and this provides a sense of our physical and physiological condition. These feelings generate the intuitive notion that bodily sensations are intrinsically tied to life, represent relevant signals for survival and well-being, and underlie mood, emotional state, and fundamental cognitive processes. (Herbert and Pollatos 2012)

Over the last decade, functional neuroscientists have discovered that the brain and gut are in almost constant communication, sending various signals back and forth to such an extent that they are found to participate in an overall, neurovisceral axis that we are selectively made aware of through the specialized sense of "*interoception.*" Originally entailing only visceral sensations arising from the inner organs, interoception represents our ability to sense the physiological conditions of all tissues of the body (Craig 2002, 2003) within an active, homeostatic monitoring (Arnhart 2011) by detecting subtle changes in the coding of body-tissue physiology, including muscles, skin, joints, and viscera (Dunn et al. 2010). Among these, visceral information is important for the physiological and behavioral regulation of the entire organism, including emotional awareness.

These internal, physiological conditions operate in a supervisory capacity through which physiological signals in the body are transmitted up to the brain, "allowing the organism to homeostatically regulate its internal state and giving rise to awareness of bodily feelings like pain, touch and temperature" (Cameron 2001) that are required for preserving health and living conditions (see Verdejo-Garcia et al. 2012). As Mayer explains, "the engagement of circuits outside of the gut wall integrates interoceptive and exteroceptive information to optimize the homeostatic regulation of intestinal function" (2011). This capacity appears most developed in primates and especially humans (Craig 2003), suggesting that it is reserved for species with a complex sense of self and social structures in tangent with the neural architecture required to constantly update values from the body's autonomic maintenance process. We reference this internal architecture under the heading of the neurovisceral axis.

More recently, attention has shifted to the generative role of active, visceral information to reach (interoceptive) awareness and influence behavior in the higher cortices of the neurovisceral axis. As Buldeo explains, "afferent information traveling from the body to the central nervous system enables the body to holistically control or maintain this homeostatic *milieu interieur*" (2015). This specifies "a ubiquitous information channel used to represent one's body from within"

(Tajadura-Jimenez and Costantini 2011) that, taken broadly, constitutes "the material me" (Craig 2003) relating to how we perceive feelings from the body, thus determining mood, sense of well-being, and emotions (see Cameron 2002), as well as decision making and behavior (Bechara and Damasio 2003; Naqvi 2010; Ernst 2014). Two initial concepts are important for understanding the roots of interoception and the brain-gut axis: *homeostasis* and the *interior milieu*.

In 1865, C. Bernard hypothesized a set of regulative centers operating for the maintenance of the body's internal survival parameters to maintain internal equilibrium, or what W. Cannon later called, homeostasis (1929). Bernard's idea was two-fold; namely, that there existed an identifiable set of physiological parameters defining the normative internal state of the organism, and that the body would seek to maintain these optimal parameters in a dynamical capacity (1865). Cannon's 1929 extrapolation of Bernard's *interior milieu* led to the concept of homeostasis as describing "a process of synchronized adjustments in the internal environment resulting in the maintenance of specific physiological variables within defined parameters including blood pressure, temperature and pH balance—all with clearly defined "normal" ranges or steady-states" (Bowden 2013). As Buldeo explains: "both the *milieu interieur* and homeostasis imply the existence in the body of mechanisms by which the organism can track the moment-to-moment fluctuations of these physiological parameters" (2015). Fifteen years after Bernard's paper, James and Lange developed the first rigorous theory of neurovisceral interactions in an emotional and behavioral capacity based on the idea that "stimuli that induce emotions such as fear, anger or love initially induce changes in visceral function through autonomic nervous system output, and that the afferent feedback of these peripheral changes to the brain is essential in the generation of specific emotional feelings" (Mayer 2011)—cf. the theory of "embodied emotion" (James 1884).

In this capacity, neurovisceral dynamics can be conceptualized as a hierarchical axis of reflex circuits spanning from the enteric nervous system of the gut up to the central nervous system of the brain. This brain-gut axis is responsible for maintaining the balance and regulation of body systems in participation with the cerebral, autonomic, and enteric nervous systems—and for generating moment-to-moment interoceptive images of the *interior milieu* (physiological homeostasis) of the human body (Damasio 2010; Craig 2003, 2009; Mayer 2011; Critchley et al. 2004) that constantly relay internal needs to subjective awareness (Craig 2002; El Aidy et al. 2014). Taken together, Cannon's homeostatic regulation of Bernard's *interior milieu* designates the arena where interoceptive awareness is commissioned through ascending afferent signals integrating with further information in the higher cortices, leading to affective values.

Within this axis, multiple signaling pathways maintain communication by facilitating interactions between gut microbiota and the central nervous system (Grenham et al. 2011; Holzer and Farzi 2014) that not only ensures the proper maintenance of gastrointestinal homeostasis and digestion, but is also likely to have multiple influences on memory formation, emotional arousal, and affective behaviors; motivation; and higher complex cognitive functions (Mayer 2011), including the specialized and adaptive capacity for intuitive decision making (Allman 2005b).

Three standard mechanisms encode sensory information in the gut: primary afferent neurons, immune cells, and entero-endocrine cells (Mayer 2011); of these, we will concern ourselves with only the first. Primary afferent fibers found throughout every aspect of the gut innervate all tissues of the body and are essential for the optimal neural regulation of visceral organs (Brozmanova et al. 2010). Specifically, Craig (2002) demonstrates how *primary afferent fibers* mediate the interoceptive axis by way of a spinothalamocortical system projecting a general path through select spinal cord and brainstem nuclei to the insular cortex where they are received in the posterior insular cortex (PIC) as interoceptive signals for further processing, until a constrained set are finally represented as affective values in the "on-flowing" stream of an individual's moment-to-moment interoceptive awareness in the anterior insular cortex (AIC).

Cortical regions reliably engaged during states of enhanced bodily arousal include the anterior cingulate and insula cortices. In humans, a re-representation of interoceptive signals arriving in the PIC is fashioned in the AIC, providing a subjective image of the material self as an emotionally aware, feeling, and sentient being (Craig 2003). In general, the AIC and ACC (anterior cingulate cortex) are both implicated in social reasoning (Allman and Watson 2007), empathy, emotion, and the monitoring of visceral activity (Allman et al 2005). Arnhart explains:

> The insular cortex receives signals from all the tissues of the body, and these signals are integrated with physical and social stimuli from outside the body and with the memory of past experiences as well as imaginative projections of future experiences. The ACC can then be activated to motivate behavior to correct whatever is wrong. This neural processing mechanism seems to be unique to primates, but it's more highly developed in human beings. (2011)

Here, the afferent fibers projected to the posterior insula (PIC) form a representation of the physiological condition of all bodily tissues (Damasio et al. 2000; Craig 2002). From here we observe a secondary operation on active information reaching the interoceptive cortex whereby, as Gu and Hof (2013) explain, a posterior-to-anterior gradient in the insula enables the physical features of interoception processed in PIC — as they work their way anterior — to then integrate with information coming from several other brain regions including those responsible for elaborating sensory information about the external world and body representations (Craig 2003). These body representations are then re-mapped onto the anterior insula where they are integrated with emotional, cognitive, and motivational information from a network of corollary brain regions. It is only at this point that emotional awareness emerges, in part, as the products of complex signal processing (see Barlassina and Newen 2014). This remapping of interoceptive signals in the AIC has been proposed to underpin not only a primary form of self-awareness, but also to participate in higher forms of self-awareness such as the distinction between self and other required for intuitive social interactions (Tajadura-Jimenez et al 2014); indeed, neuroimaging studies consistently show differential activation of the AIC associated with intuition (Kuo et al. 2009). We now turn to consider the evolutionary role of specialized, von Economo neurons, which seem to have evolved in order to facilitate the rapid processing of information.

In light of the last section, we discover that the neurovisceral axis enables a modelling of intuitive decision making based on the representation of affective markers (Damasio 2010) predicated on gut responses working in concert with the evolutionarily rare and morphologically specialized *von Economo neurons* (VENs), found only in species with a highly complex social structure and sense of self—and most pronounced by far in the human brain (Allman, Watson and Tetreault 2005; Watson et al. 2006). From this perspective, and based on laws of signal conductance, VENs were likely adapted to synthesize and rapidly transmit vast amounts of information into simple signals distributed profusely throughout the brain (Watson 2006) that enable fast-paced (intuitive) decision making based on interoceptive stimuli (Preuschoff 2008), often involving social information. As Allman explains:

> Intuition is a form of cognition in which many variables are rapidly evaluated in parallel and compressed into a single dimension. This compression facilitates fast decision-making. Typically we are not aware of the logical steps or assumptions underlying this process although intuition is based on experience-dependent probabilistic models. Instead we experience the intuitive process viscerally. Intuition operates largely in the social domain but can also be applied to purely physical situations [as in physics]. Intuition is plastic; it is not instinct, although instinctive feelings may contribute to it. Emotional value judgments contribute to both intuition and deliberation. (2005b)

In a morphological and functional capacity, VENs likely represent a specialization of pyramidal cells where all the growth is concentrated in a single, massive dendrite similar to a motor projection neuron; as Allman explains, "this specialization enables us to reduce complex social and cultural dimensions of decision-making into a single dimension that facilitates the rapid execution of decisions" (2005b). VENs appear late in both evolution and ontogeny, arising around 15,000 and 30,000 years ago in a convergent evolution of 1) pachyderms and cetaceans, and 2) in the great apes/hominoid line—and only arising postnatally in the first few years of life (Allman, 2002). In addition, VENs are only found in three regions of the brain: the anterior insular cortex (AIC), anterior cingulate cortex (ACC), and dorsolateral prefrontal cortex (dlPFC). Differential activity of the dlPFC is correlated with deliberative cognition whereas AIC/ACC are more emotional (Fajardo, 2008), indicating that intuitive processing sends the same signals to deliberative—as well as emotional—centers of the brain. Within these regions, VENs only appear in the fifth laminar layer of a six-layered laminar structure of the cortex, which classically designates an output layer, thereby reinforcing the idea that VENs are involved in the diffuse projection of information throughout the brain. Indeed, VENs proliferated *greatly* in the human line: of the roughly 87 billion neurons in the adult human brain, VENs in the AIC number (on average) upwards of 193,000 in adult humans while in the brains of centenarians those numbers shoot into the range of 500,000 and beyond. By comparison, in one of the nearest genetic relatives, the gorilla, the average volume-count of VENs dramatically diminishes to a mere 16,500 (Allman, 2005a). In addition, there have been found roughly 30 percent more neurons in the right hemisphere than the left hemisphere of AIC, suggesting a functional specialization involving social elements of information processing (Butti et al. 2010).

To scaffold these claims, VENs demonstrate a neurotransmitter receptor signature that specifically indicates a role in rapidly sending and receiving signals from the gut (Allman 2009, 2011; Watson et al. 2006; Allman and Watson 2007). In terms of a molecular profile, VENs strongly express dopamine—a high-affinity receptor that is expressed *only* on VENs. Serotonin 2b also has receptors on VENs. This is interesting because these two receptors are very rarely seen in the central nervous system, but instead in the mesenteric (peripheral) nervous system, where they regulate gastric motility, transforming into visceral feelings. In addition, VENs have been shown to express neuromedin B, which is also involved in gastric motility. The strong expression of all three of these receptors in VENs ardently implicates a relationship with the gut. A fourth neurotransmitter strongly expressed on VENs is the vasopressin 1a receptor, necessary for mediating the formation of social bonds.

This evidence provides further support for the functional hypothesis presenting a modulatory and mediating role for VENs in the swift transmission of visceral afferents from the gut to the brain—and throughout the brain, as adapted for the fast-acting consideration of, and response to, vastly complex degrees of social information that the gut can process on a level far superior to cognition, that is, intuition-like. Taken in this capacity, the von Economo neurons confer a multi-scale assessment of social situations by virtue of a rapid signal processing transmitted diffusely throughout the brain to enable intuitive feelings designed to inform the attention, assessment and, if necessary, an allostatic behavioral response during socially complex situations and complex decision making and problem solving. Allman provides the authoritative assessment:

> Von Economo neurons may relay a vast intuitive assessment of complex social situations to facilitate the rapid adjustment of behavior in quickly changing social situations. [ . . . ] We hypothesize that this specialization enables us to reduce complex social and cultural dimensions of decision-making into a single dimension that facilitates the rapid execution of decisions. Other animals are not encumbered by such elaborate social and cultural contingencies to their decision making and thus do not require such a system for rapid intuitive choice. (2005b)

Specialized in every capacity, VENs have an important role to play in intuition, enabling us to overcome uncertainty, make quick decisions, and resolve cognitive dissonance (Allman 2002, 2005a).

## RETRACING INTEROCEPTION AND NEUROVISCERAL AXIS IN WHITEHEAD

Taking the preceding premises, we can now apply them to Whitehead's work to demonstrate a general foresight into interoception and the regulatory capacity of neurovisceral dynamics through his conceptual distinction between consciousness and subjective awareness (PR 36, 112). Such a distinction is further compatible with Gerald Edelman's "primary consciousness" vs. "higher-ordered consciousness," and Antonio Damasio's "proto-self" vs. "autobiographical self," though we will only name them here. Secondly, this distinction is articulated in Whitehead's theory of perception in the three modes of causal efficacy, presentational immediacy, and symbolic reference. Of these, the first two modes will prove to align remarkably well with neurovisceral dynamics and interoception. In both cases, Whitehead shows a prescient awareness of the underwriting autopoietic architecture and dynamics in the brain/body before they were specified within neurobiology. Thirdly, reexamining the neurovisceral axis, we immediately recognize the role that "primary afferents" fulfill qua Whitehead's initial feelings (PR 236, 362). Locating this starting position, we build up through Whitehead's prehension and concrescence to satisfaction.

## CONSCIOUSNESS VS. SUBJECTIVITY

Whitehead esteemed the notion that "the original locus of perception is not in consciousness but in the organism's manifold forms of autopoiesis . . . necessarily congruent with the environment in which it manages to survive," as Weekes (2010) describes. Here we can identify autopoiesis with the homeostatic, regulatory capacity engaged in interoception. Echoing this perspective, Weekes supplements the notion of a primitive experience as a "physical process at its most elementary level: a faint wisp of affective experience 'enjoyed' by an extremely rudimentary nonconscious subjectivity." Applying this language, we easily recover the sense in which a constrained element of total signal processing is integrated in the middle insular cortex (MIC), and re-represented in the AIC as affective/emotional feeling qua interoceptive awareness. This 'nonconscious subjectivity' (Weekes 2010) is appropriately represented by the regulatory principle of homeostasis and the representational capacity in the insula; as Mayer explains: "The conscious awareness of interoceptive images that is generated either by signals from the gut or by the recall of interoceptive memories of such signals is associated with the perception of emotional feelings, including pain, disgust or well-being" (2011).

In several instances, Whitehead identifies affective and emotional experiences as the most basic process that does not, at its most rudimentary level, involve consciousness, thought, or sense-perception[1] (PR 36) — applying only to higher and later phases of concrescence in more complex actual entities (Marstaller 2009). For example: "the basic fact is the rise of an affective tone originating from things whose relevance is given" (AI 226). This can also be compared to a rich continuity of thought in neurophenomenological views, citing in one case how "affect and emotions [reflect] the originary source of the living present" (Varela and Depraz 2005), and in another case, describing our "primitive self-consciousness [as] fundamentally linked to bodily processes of life regulation, emotion and affect, such that all cognition and intentional action are emotive" (Zelazo et al. 2007).

Reading these descriptions against the backdrop of autopoietic, interoceptive dynamics, this type of activity is demonstrably covalent with ascending signals from the primary afferent fibers mediating interoceptive channels (qua active information) of the body's *interior milieu* in order to selectively achieve awareness in the AIC. To these ends, we acquire an undeniably-rich concomitance between interoception and the neurovisceral axis defined in the context of homeostatic regulation, and as a basis for affective awareness. Without exploring it here, we find further evidence for such a primordial program underwriting consciousness in three other places: Edelman's "primary consciousness" (2004); Damasio's "proto-self" (2003); and Craig's "material me" (2003).

WHITEHEAD'S THEORY OF PERCEPTION

The basis for an autopoietic, regulative capacity in the body, as primary subjectivity, bears a remarkable connection to Whitehead's "perception in the mode of causal efficacy" as one of three primary modes of perception, also including *presentational immediacy* and *symbolic reference* (Hooper et al. 1944; Maclachlan 1992; Riffert 2004). Whitehead describes perception in the mode of *presentational immediacy* as representing an awareness of sense-data "by means of our projections of our immediate sensations, determining for us characteristics of contemporary physical entities" (Whitehead 1927). Here we can correlate "sense-data" with interoceptive signals in PIC, serving as "a world decorated by sense-data dependent on the immediate states of relevant parts of our own bodies" (*ibid*). We account for these relevant parts as two primary pathways in the neurovisceral axis: (1) through afferent fibers leading to the insula; and (2) through skin afferents leading to the multimodal,

somatosensory cortex yielding proprioceptive, exteroceptive, and interoceptive awareness (see Mayer 2011).

Whitehead further describes presentational immediacy as "an
outgrowth from the complex datum implanted by causal efficacy"
(PR 173), which in this case can be referred to the more primordial,
luminal basis of the perception of the interior milieu. For a direct
link to interoception we turn to Weekes' description of *causal efficacy*
as "mostly visceral in locus and affective in content" (2010), further
explaining how "for Whitehead, causal efficacy is the bodily feeling
we have of being affected by and dependent on an ambient physical
milieu, as well as the feeling we have of the organic body itself as the
most proximate milieu of this sort" (*ibid*). Following suit, Cromby's
description can be read into a Jamesian discourse, explaining how "the
moment-by-moment flow of our experience consists, before it consists
of anything else, of a flow of embodied sensations or feelings (Cromby
2006). Weekes continues:

> The organism adapts itself moment by moment to ambient
> influences from its immediate environment. This happens in
> internal processes of homeostasis that continually compen-
> sate for fluctuations in critical parameters of the organism's
> milieu, as well as in reflex movement and the many forms of
> arousal and affectivity that are elicited biochemically and do
> not depend on conscious mediation. (2010)

Even without technical acquaintance, Weekes' description provides a
precise harbinger of interoceptive properties through the identification
of two main components: through feeling-values as visceral afferents,
and through the end product of processing in higher cortices as represented in the mode of affective and emotional values. In another capacity, Weekes describes how perception in the mode of causal efficacy is
"characteristically vector-like in conveying vague but imperative information" (2010). We apply this to the vector-like nature of ascending
afferent fibers to the PIC.

Combining these two modes of perception we consider how "the
function of the phase of presentational immediacy is to provide a representation or mapping of the datum, which is dimly discerned at the

level of causal efficacy" (Maclachlan 1992). In this sense, *presentational immediacy* refers to the ongoing representation of values in the primary somatosensory area (Craig 2003) and as the final, interoceptive afferents expressed in the AIC: the capstones of concrescent values synthesized in the neurovisceral axis. Further, read against current scholarship, Whitehead's perception in the mode of *causal efficacy* effectively earmarks what has since come to be understood thru the lens of regulative dynamics communicating through primary afferent fibers underwriting the brain-gut axis and finally expressed as "interoceptive awareness" qua affect/emotion in the AIC.

PREHENSION, CONCRESCENCE AND SATISFACTION
IN NEUROVISCERAL (NV) DYNAMICS

Next, the link between Whitehead's prehension and concrescence and the neurovisceral axis is straightforward in correspondence. We consider *prehension* to refer to the analysis of components that arise from their physical data — and *concrescence* to the way in which positive prehensions creatively coalesce to form the *satisfaction* of that actual occasion. Marstaller's explanation provides nice accompaniment:

> A prehension is an active relation between two actual entities.
> It is directed from the past to the new entity. Both are therefore
> called object and subject, respectively. Its activity is its taking
> part in concrescence. (2009)

Coordinating this description in the neurovisceral axis, rather than "subject" and "object" we consider prehensions as the projection of affective fibers combining with others as they relay from node-to-node up the axis. In its original context, as Whitehead explains, *concrescence* is a derivative from the Latin verb meaning to "grow together" (AI 303). Taken as such, "an actual occasion is a concrescence effected by a process of feelings" (PR 211). The driving operation during the phases of concrescence is found to bear in the analysis of prehensions. Whitehead describes how "the first analysis of an actual entity into its most concrete elements discloses it to be a concrescence of prehensions which have originated in its process of becoming" (PR 35). In an interoceptive context, primary afferent fibers mediate the brain-gut axis, relaying visceral feelings of the body-interior, or what Whitehead calls "initial feelings," in a "concrescual" capacity (Pred 2009) up to the higher cortices, and arriving in the PIC as what Whitehead calls the "presentational immediacy" of a "multiplicity of initial data":

> A multiplicity of simple physical feelings . . . constitutes the first phase in the concrescence of the actual entity which is the common subject of all these feelings. (PR 236)

> All the more complex kinds of physical feelings arise in subsequent phases of concrescence, in virtue of integrations of simple physical feelings with each other and with conceptual feelings. (PR 245)

Here, "simple physical feelings" correspond with afferent feelings in the neurovisceral axis, and a multiplicity of these feelings in the PIC constitute the initial phase of concrescence in the generative process of an AE. Subsequent phases are constituted by a succession of "more complex feelings integrating the earlier simpler feelings, up to the satisfaction which is one complex unity of feeling" (PR 220). These subsequent phases of concrescence can be correlated with the dynamics of insular processing where values are subject to a progressive integration with other prehensions before concrescing into a final re-representational value in the AIC, or *satisfaction*, that confers affectively as "interoceptive awareness." To buffer this claim, Pred's concrescual (*Onflow*) dynamics are applied to the regulative monitoring network of the *interior milieu* where Pred considers the origin of values in concrescence in terms of visceral, regulatory dispositions triggering hedonic, endocrine, and other set points — in other words, of interoception and the maintenance of homeostasis within internal systems; only, instead of "set points" we acknowledge a dynamic range of optimal states. This effectively renders the entire axis concrescual.

For the sake of consistency we adopt Whitehead's naming convention to recognize "initial feelings" as "primary afferents" in the NV axis, and "prehensions" as those values during the predictive coding phase (Seth 2013) in the insula that yield positive or negative values. The values that emerge from this coding for the future phases of concrescence are the positive prehensions; thus, to be positively prehended is to become a feeling of a new subject for the later phases of concrescence (Sherburne 1966). This makes concrescence a "concrescence of prehensions" (PR 35) whose goal is to resolve into a final satisfaction in the mode of an affective/emotional representation constituting interoceptive awareness.

Putting the two descriptions together, Whitehead's "feelings"—as primary afferent fibers—ascend the neurovisceral axis to the PIC as "a multiplicity of data" (PR 212; Ford 1984), all the while participating in an ongoing concrescence with other feelings gleaned in the process. The axis itself represents a multi-level distribution of internal regulative centers between the luminal environment of the enteric nervous system and the higher cortices of the central nervous system. As such, "afferent feelings" ascending a concrescual, multi-scale neurovisceral axis represent the active information-dynamics of interoception. This distinction provides insight and an update for signal processing; more specifically, it updates the characterization of bottom-up models of perception from a simple, additive accumulation to a selective, concrescual logic detailing how prehensions are constantly "growing together" (AI 303) in a selective, organic synthesis. As Marstaller explains:

> Because of being organic, a concrescence is different from a mere combination where the whole is the sum of its parts. . . . The concrescence of an actual entity constitutes a process of integrating the multiple feelings, or "prehensions", into one subjective/superjective unity. (2009)

Taken as such, Whitehead's phases of concrescence represent an operational phase of the generative process whereby positively prehended feelings organize into optimal combinations of "conformal " and "comparative feelings" capable of reordering with each successive phase of concrescence and the new data that are introduced. When we apply this to insular processing dynamics we recognize the further information integrated into the MIC and AIC as salient and cognitive information signals processing anteriorly where they are expressed as affective, emotional value-representations.

This leaves the final stage: satisfaction, as the delta where signals finally converge into re-representational form as an affective/emotional value in the AIC; or in Whitehead's description, as the final phase of concrescence where prehensions are integrated into a concrete unity, or feeling (PR 38). Similarly, Sherburne highlights Whitehead's definition of concrescence as "the process of integrating the initial welter of many simple physical feelings into the one complex unity of feeling that is the satisfaction" (41). In any event, "it is clear that the satisfaction is a feeling" (Christian 25). Indeed, Whitehead is shown to have described this in numerous passages: (e.g., PR 38-9; 66; 71; 434; AI 298).

In another instance, Whitehead describes a "satisfaction" as "the concrete unity of feeling" obtained by the process of integration (PR 322). Here we consider "integration" in terms of the integration of signals from the hypothalamus, prefrontal cortex, amygdala, and ACC with signals in the MIC and posterior AIC, just prior to the final representation of value, or "satisfaction," in the AIC. Further support for this type of posterior-to-anterior gradient processing and progressive integration in the insula is conceptually identified by Sherburne, explaining how concrescence is divisible into an initial stage of many feelings and a subsequent succession of phases of more complex feelings that integrate the earlier, simpler feelings until achieving the final "satisfaction" as a "complex unity of feeling" (44). This makes it easy to imagine the succession of phases in terms of the posterior-to-anterior processing gradient (Gu and Hof et al. 2013; Cacioppo et al. 2013; Craig 2014) during which time more complex information is integrated from other networked brain-regions to provide further scaffolding for the re-representations of interoceptive feelings in AIC.

Whitehead further explains how a *satisfaction* is an experience that has "intensity" (PR 129), or more specifically, a "quantitative emotional intensity" (177) that is "subjective" (82) and "immediately felt" (235). Combining these references we recover a strong sense for the salience (intensity) ascribed to certain interoceptive signals underwriting the registry of awareness as affective/emotional, and often registering for the sake of precipitating an allostatic response from the individual. Taken in Whitehead's sense, the proportional salience ascribed to certain values strongly resembles the prehension process that selectively identifies values for the further phases of concrescence in the MIC before final re-representation as an affective/emotional "satisfaction" (or 'concern') in the AIC qua interoceptive awareness.

Diagrammatically, we envision the logic to proceed in the following "geomodal" (Bettinger 2015) organization where initial feelings, like primary afferents, generally ascend to PIC as a multiplicity of data (mD). These data are prehended and positive values go into the phases of concrescence in the MIC where they are integrated with other information until forming a final value, or satisfaction (SAT), in the AIC.

All told, it is somewhat remarkable that Whitehead's insights *not*

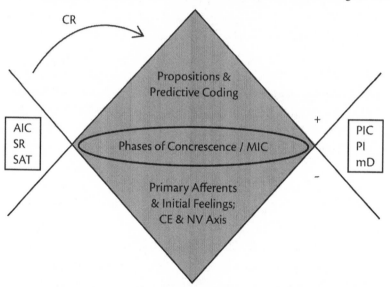

*only* serve to "correct" philosophy by refocusing it on the most fundamental, affective predicates at the subconscious level operating in autopoietic dynamics inhering in subjective-awareness (sense of self) underwriting experience, but also that this can be linked to the homeostatic, regulatory maintenance of the body. This serves as a primitive layer of being, for Whitehead, in the mode of affect/feeling—or in the mode of interoceptive afferents concrescually contextualizing through the brain-gut axis into emotional awareness. In so doing Whitehead effectively provides a nascent description for what has come to be qualified under the heading of interoception. The conceptual richness of

Whitehead's framework is once again exemplified in this application, just as it has also been found to provide insightful mappings to key elements of quantum field theory and string theory (Bettinger, 2015). Indeed, renormalization dynamics and information processing in the insula — qua interoception and emotional awareness — suggest several motivational overlaps to be developed in a future paper.

Revisiting our initial questions, we find that Whitehead's theory of perception and the causal dynamics of AEs finds a surprisingly precise concomitance within the modern development of interoception, interoceptive awareness, and the brain-gut axis. Secondly, we can leverage Whitehead's pure philosophy to provide reciprocal insight into insular dynamics underwriting interoceptive awareness to the extent that precise predictions can be made in follow-up works regarding more technical components of interoception. In fact, the more technically we venture, the more alignment appears. Indeed, this brief overview is just the beginning.

## SUMMARY AND CONCLUSION

> Speculative boldness must be balanced by complete humility before logic, and before fact. (PR 25)

Contemporary developments in neuroscience and microbiology indicate a rigorous basis for the interoceptive faculty and microbial brain-gut axis to engage a neurobiological account of intuition following similar lines of description mimicking the dynamics of recently successful theories in physics calling for active information integration in scale-free and multi-level macrosystems (see Eastman 2004). In this likeness, the brain-gut axis refers to an unparalleled relationship of active multi-scale, reflexively communicating interoceptive systems between the gastrointestinal tract and the brain, underwriting homeostasis and higher-complexity cognitive functions including social dynamics and intuitive decision making. To these ends, interoception represents the way in which we are acquainted with this axis and are ever responsive to and aware of the patently personal, internal environment of our bodies. To scaffold this claim, Dunn et al. explain how "emotion, experience and intuition are associated with individual differences in the ability both to generate and to perceive accurately subtle changes in the body" (2010). As such, these processes operate uniquely in response to each individual.

In concert with the neurovisceral axis, the biological basis for intuition is greatly enabled by the existence of the evolutionarily-specialized von Economo neurons, only existing in the AIC, ACC and dlPFC of species with highly complex social structures and a highly developed sense of self. These neurons are evolutionarily tailored to synthesize and transmit vast amounts of information into simple signals distributed profusely throughout the brain for fast-acting responses leading to adaptive and intuitive results. Applying Whitehead's dynamics to interoception we are able to distinguish a further point: namely, that bottom-up dynamics are more than just an aggregation of sensory signals; rather, an application of Whitehead's framework, further reinforced by Pred, suggests that afferent signals more-accurately "concresce"—that is, they "grow together" (AI 303) as opposed to merely accruing. This implies a more complex dynamical picture than is currently perceived in neuroscience.

In a grander sense, Whitehead is shown to have largely (if not categorically) predicted what has come to be identified under the heading of interoception and the affective nature of neurovisceral afferents (Craig 2003, 2009; Mayer 2011) as the information mediators of this axis. As indicated here, this is most evident in three capacities: (1) in terms of Whitehead's general distinction between experience and subjectivity; (2) in terms of his identification of perception in the modes of *causal efficacy* and *presentational immediacy*; and (3) as a correlation of primary causality between neurovisceral dynamics and Whitehead's causal dynamics entailed in *prehension, concrescence,* and *satisfaction*. Here, afferent signals built up through the interoceptive axis to the PIC integrate with other values in the MIC before re-representation in the AIC as interoceptive awareness; thus, interoceptive afferents only reach the level of sensory and emotional awareness in the AIC. Putting these models together we render the following diagrammatic logic:

This model correlates the dynamics of Whitehead's theory of

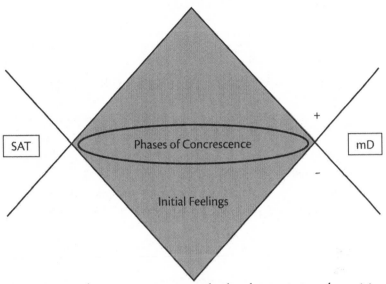

perception and AEs in-concert-with the dynamics underwriting interoception and the "concrescual" NV axis. Following the model, Whitehead's "*initial feelings*" in the mode of causal efficacy (CE) correlate with 'primary afferents' in the neurovisceral axis. These values project through spinothalamocortical pathways to the posterior insular cortex (PIC) as the "presentational immediacy" (PI) of a *multiplicity of initial data* (mD), for Whitehead, or as visceral, interoceptive signals. This multiplicity of data is then *prehended* to determine positive or negative values (+/-) of which only the positive values proceed into the further *phases of concrescence* for integration with a welter of other information. This phase corresponds with the posterior-to-anterior processing and integration of information in the insula, locating much of the integration in the MIC. All this processing of positive prehensions culminates in a final, affective representation as a "satisfaction," for Whitehead, or as interoceptive awareness in the AIC. Given the values in interoceptive awareness are those that become consciously perceived, this correlates with Whitehead's perception qua *symbolic reference* (SR) as linked to consciousness.

This model also foretells some of the further reaches not able to be explored in this essay in the interest of space, but that deserve some mention in closing. Mainly, the top-down aspect of interoception still stands to be introduced in the capacity of inferential, predictive coding, and Bayesian brain dynamics (Seth 2013). Such an approach acquaints us with the "free-energy principle" (Friston 2006): a concept with roots in Freudian theory and thermodynamics as well as in Feynman's *path integral* notation and, as we will see, in critical components of Whitehead's AE dynamics. To leave you with one of the major connections, it can be shown that "creativity" and "novelty" are precisely the products of abductive processing in the context of interoceptive awareness. All told, it seems we may have finally found a rigorous biological basis for Whitehead's organic philosophy within the context of interoception and the neurovisceral axis underwriting consciousness. The more we explore this axis, the more we will recover critical elements of Whitehead's scholarship.

## NOTES

1    Identifying "sense-perception" with the standard five senses, we maintain (in this context) the distinction between sense-perception and interoception (the sense of the inner environment of one's body).

## REFERENCES

Allman, John, Atiya Hakeem, and Karli Watson. "Book review: Two Phylogenetic Specializations in the Human Brain." *The Neuroscientist* 8.4 (2002): 335–46.

Allman, John M., Karli K. Watson, Nicole A. Tetreault, and Atiya Y. Hakeem. "Intuition and Autism: A Possible Role for Von Economo Neurons." *Trends in Cognitive Sciences* 9.8 (2005): 367–73.

Allman, John; "Brain, Mind and Consciousness." Presentation at Skeptics Conference, Pasadena, CA; 13–15 May, 2005.

Allman, John M., Nicole A. Tetreault, Atiya Y. Hakeem, Kebreten F. Manaye, Katerina Semendeferi, Joseph M. Erwin, Soyoung Park, Virginie Goubert, and Patrick R. Hof. "The von Economo Neurons in Frontoinsular and Anterior Cingulate Cortex in Great Apes and Humans." *Brain Structure and Function* 214.5–6 (2010): 495–517.

Allman, John, and Karli Watson. "Neuroanatomy of the von Economo Neurons, a Recently Evolved Neuronal Population in Great Apes and Humans." In *The Evolution of Primate Nervous Systems*, ed. J. H. Kaas and T. M. Preuss. Oxford: Elsevier, 2007.

Arnhart, Larry. "Darwinian Conservatism." (Blog Comment of June 8, 2011: http://darwinianconservatism.blogspot.com/2011/06/bud-craig-on-interoception-and.html.)

Barlassina, Luca, and Albert Newen. "The Role of Bodily Perception in Emotion: In Defense of an Impure Somatic Theory." *Philosophy and Phenomenological Research* 89.3 (2014): 637–78.

Bechara, Antoine, Hanna Damasio, and Antonio R. Damasio. "Role of the Amygdala in Decision-making." *Annals of the New York Academy of Sciences* 985.1 (2003): 356–69.

Bernard, Claude. *Introduction à l'étude de la médecine expérimentale.* Baillière, 1865.

Bettinger, Jesse Sterling, "The Founding of an Event-Ontology: Verlinde's Emergent Gravity and Whitehead's Actual Entities." Order No. 3701113: The Claremont Graduate University, 2015 (http://search.proquest.com/docview/1680842524?accountid=131239).

Bowden, Mark. *Tame the Primitive Brain: 28 Ways in 28 Days to Manage the Most Impulsive Behaviors at Work.* New York: John Wiley & Sons, 2013.

Kollarik, M., F. Ru, and M. Brozmanova. "Vagal Afferent Nerves with the Properties of Nociceptors." *Autonomic Neuroscience* 153.1 (2010): 12-20.

Buldeo, Nitasha. "Interoception: A Measure of Embodiment or Attention?" *International Body Psychotherapy Journal* 14.1 (2015).

Bullock, Theodore, and G. Adrian Horridge. "Structure and Function in the Nervous Systems of Invertebrates." New York: W.H. Freeman, 1965.

Butti, Camilla, Chet C. Sherwood, Atiya Y. Hakeem, John M. Allman, and Patrick R. Hof. "Total Number and Volume of Von Economo Neurons in the Cerebral Cortex of Cetaceans." *Journal of Comparative Neurology* 515.2 (2009): 243–59.

Norman, Greg J., Louise Hawkley, Aaron Ball, Gary G. Berntson, and John T. Cacioppo. "Perceived Social Isolation Moderates the Relationship between Early Childhood Trauma and Pulse Pressure in Older Adults." *International Journal of Psychophysiology* 88.3 (2013): 334–38.

Cameron, Oliver G. "Interoception: The Inside Story—A Model for Psychosomatic Processes." *Psychosomatic Medicine* 63.5 (2001): 697–710.

Cameron, Oliver G., and Satoshi Minoshima. "Regional Brain Activation Due to Pharmacologically Induced Adrenergic Interoceptive Stimulation in Humans." *Psychosomatic Medicine* 64.6 (2002): 851-61.

Cannon, Walter B. "Organization for Physiological Homeostasis." *Physiological Reviews* 9.3 (1929).

Christian, William A. *An Interpretation of Whitehead's Metaphysics.* New Haven, Yale Univ. Press, 1959.

Craig, A. D. "An Ascending General Homeostatic Afferent Pathway Originating in Lamina I." *Progress in Brain Research* 107 (1995): 225-42.

Craig, Arthur D. "How Do You Feel? Interoception: The Sense of the Physiological Condition of the Body." *Nature Reviews Neuroscience* 3.8 (2002): 655-66.

Craig, A. D. "Interoception: The Sense of the Physiological Condition of the Body." *Current Opinion in Neurobiology* 13.4 (2003): 500-05.

Craig, A. D. "How Do You Feel — Now? The Anterior Insula and Human Awareness." *Nature Reviews Neuroscience* 10.1 (2009): 59-70.

Craig, A. D. *How Do You Feel? An Interoceptive Moment With Your Neurobiological Self.* New Jersey: Princeton University Press, 2015.

Critchley, Hugo D., Stefan Wiens, Pia Rotshtein, Arne Öhman, and Raymond J. Dolan. "Neural Systems Supporting Interoceptive Awareness." *Nature Neuroscience* 7.2 (2004): 189-95.

Cromby, John. "Reconstructing the Person." © British Psychological Society, Clinical Psychology Forum 162 (2006): 13-16.

Damasio, A. R. *The Feeling of What Happens: Body and Emotion in the Making of Consciousness.* New York: Harcourt Brace, 1999.

Damasio, Antonio R., Thomas J. Grabowski, Antoine Bechara, Hanna Damasio, Laura L. B. Ponto, Josef Parvizi, and Richard D. Hichwa. "Subcortical and Cortical Brain Activity During the Feeling of Self-Generated Emotions." *Nature Neuroscience* 3.10 (2000): 1049-56.

Damasio, Antonio. "Mental Self: The Person Within." *Nature* 423.6937 (2003): 227-27.

Damasio, Antonio. *Self Comes to Mind: Constructing the Conscious Brain.* New York: Pantheon, 2010.

Dunn, Barnaby D., Hannah C. Galton, Ruth Morgan, Davy Evans, Clare Oliver, Marcel Meyer, Rhodri Cusack, Andrew D. Lawrence, and Tim Dalgleish. "Listening to Your Heart: How Interoception Shapes Emotion Experience and Intuitive Decision Making." *Psychological Science* (2010): 1835-44.

Eastman, Timothy E., and Hank Keeton, eds. *Physics and Whitehead: Quantum, Process, and Experience.* SUNY Press, 2004.

Eastman, Timothy E. "Our Cosmos, From Substance to Process." *World*

*Futures* 64.2 (2008): 84-93.

Edelman, Gerald M. *Wider Than the Sky: The Phenomenal Gift of Consciousness.* New Haven: Yale University Press, 2004.

El Aidy, Sahar, Timothy G. Dinan, and John F. Cryan. "Immune Modulation of the Brain-Gut-Microbe Axis." *Frontiers in Microbiology* 5 (2014).

Ernst, Jutta, Heinz Böker, Joe Hättenschwiler, Daniel Schüpbach, Georg Northoff, Erich Seifritz, and Simone Grimm. "The Association of Interoceptive Awareness and Alexithymia with Neurotransmitter Concentrations in Insula and Anterior Cingulate." *Social Cognitive and Affective Neuroscience* (2013): 857-63.

Fajardo, Camilo, Martha Isabel Escobar, Efraín Buriticá, Gabriel Arteaga, John Umbarila, Manuel F. Casanova, and Hernán Pimienta. "Von Economo Neurons are Present in the Dorsolateral (Dysgranular) Prefrontal Cortex of Humans." *Neuroscience Letters* 435. 3 (2008): 215-18.

Friston, Karl, James Kilner, and Lee Harrison. "A Free Energy Principle for the Brain." *Journal of Physiology-Paris* 100.1 (2006): 70-87.

Garfinkel, Sarah N., Adam B. Barrett, Ludovico Minati, Raymond J. Dolan, Anil K. Seth, and Hugo D. Critchley. "What the Heart Forgets: Cardiac Timing Influences Memory for Words and is Modulated by Metacognition and Interoceptive Sensitivity." *Psychophysiology* 50.6 (2013): 505-12.

Grenham, Sue, Gerard Clarke, John F. Cryan, and Timothy G. Dinan. "Brain-Gut-Microbe Communication in Health and Disease." *Frontiers in Physiology* 2 (2011): 94.

Gu, Xiaosi, Patrick R. Hof, Karl J. Friston, and Jin Fan. "Anterior Insular Cortex and Emotional Awareness." *Journal of Comparative Neurology* 521.15 (2013): 3371-88.

Hartshorne, Charles. *The Philosophy and Psychology of Sensation.* Port Washington, New York: Kennikat Press, 1934.

Herbert, Beate M., and Olga Pollatos. "The Body in the Mind: On the Relationship Between Interoception and Embodiment." *Topics in Cognitive Science* 4.4 (2012): 692-704.

Holzer, Peter, and Aitak Farzi. "Neuropeptides and the Microbiota-Gut-Brain Axis." In *Microbial Endocrinology: The Microbiota-Gut-Brain Axis in Health and Disease*, 195-219. New York: Springer, 2014.

Hooper, Sydney E. "Whitehead's Philosophy: Theory of Perception" *Philosophy* 19.73 (1944): 136-58.

James, William. "What is an Emotion?" *Mind* 9 (1884): 188–205.

Johnson, A. H. "The Psychology of Alfred North Whitehead." *The Journal of General Psychology* 32.2 (1945): 175-212.

Khalsa, Sahib S., David Rudrauf, Antonio R. Damasio, Richard J. Davidson, Antoine Lutz, and Daniel Tranel. "Interoceptive Awareness in Experienced Meditators." *Psychophysiology* 45.4 (2008): 671-77.

Khalsa, Sahib S., David Rudrauf, and Daniel Tranel. "Interoceptive Awareness Declines with Age." *Psychophysiology* 46.6 (2009): 1130-36.

Kuo, Bo-Cheng, Anling Rao, Jöran Lepsien, and Anna Christina Nobre. "Searching for Targets Within the Spatial Layout of Visual Short-Term Memory." *The Journal of Neuroscience* 29.25 (2009): 8032-38.

Langer, Susanne. *Mind: An Essay on Human Feeling.* Baltimore, Maryland: The John Hopkins University Press, 1988.

Maclachlan, D. L. C. "Whitehead's Theory of Perception." *Process Studies* 21.4 (1992): 227-30.

Marstaller, Lars. "Towards a Whiteheadian Neurophenomenology." *Concrescence* 10 (2009).

Mayer, Emeran A. "Gut Feelings: the Emerging Biology of Gut-Brain Communication." *Nature Reviews Neuroscience* 12.8 (2011): 453-66.

Moreno-López, Laura, Emmanuel Andreas Stamatakis, María José Fernández-Serrano, Manuel Gómez-Río, Antonio Rodríguez-Fernández, Miguel Pérez-García, and Antonio Verdejo-García. "Neural Correlates of Hot and Cold Executive Functions in Polysubstance Addiction: Association Between Neuropsychological Performance and Resting Brain Metabolism as Measured by Positron Emission Tomography." *Psychiatry Research: Neuroimaging* 203.2 (2012): 214-21.

Naqvi, Ammar, Huzefa Rangwala, Ali Keshavarzian, and Patrick Gillevet. "Network-Based Modeling of the Human Gut Microbiome." *Chemistry & Biodiversity* 7.5 (2010): 1040-50.

Nimchinsky, Esther A., Brent A. Vogt, John H. Morrison, and Patrick R. Hof. "Spindle Neurons of the Human Anterior Cingulate Cortex." *Journal of Comparative Neurology* 355.1 (1995): 27-37.

Northoff, Georg, "The Intuition of the Self / Brainstorm and Neurons: Intuition, Creativity and Imagination." Tower of Senses Symposium, Nürnberg, Germany, 2009.

Pred, Ralph. *Onflow: Dynamics of Consciousness and Experience.* Boston, Massachusetts: MIT Press, 2005.

Preuschoff, Kerstin, Steven R. Quartz, and Peter Bossaerts. "Human

Insula Activation Reflects Risk Prediction Errors as Well as Risk." *The Journal of Neuroscience* 28.11 (2008): 2745-52.

Riffert, Franz G. "Whitehead's Process Philosophy as Scientific Metaphysics." In *Physics and Whitehead: Quantum, Process, and Experience,* edited by Timothy E. Eastman and Hank Keeton, 199-222. Albany, NY: State University of New York Press, 2004.

Riffert, F. and Michel Weber, eds. *Searching for New Contrasts: Whiteheadian Contributions to Contemporary Challenges in Neurophysiology, Psychology, Psychotherapy and the Philosophy of Mind.* Franfurt am Main: Peter Lang, 2003.

Seeley, William W., Florian T. Merkle, Stephanie E. Gaus, John M. Allman, Patrick R. Hof, and C. V. Economo. "Distinctive Neurons of the Anterior Cingulate and Frontoinsular Cortex: A Historical Perspective." *Cerebral Cortex* 22.2 (2012): 245-50.

Seth, Anil K., and Hugo D. Critchley. "Extending Predictive Processing to the Body: Emotion as Interoceptive Inference." *Behavioral and Brain Sciences* 36.3 (2013): 227-28.

Sherburne, Donald W., ed. *A Key to Whitehead's Process and Reality.* Bloomington, IN: Indiana University Press, 1966.

Tajadura-Jiménez, Ana, and Manos Tsakiris. "Balancing the 'Inner' and the 'Outer' Self: Interoceptive Sensitivity Modulates Self–Other Boundaries." *Journal of Experimental Psychology: General* 143.2 (2014): 736.

Tsakiris, Manos, Ana Tajadura-Jiménez, and Marcello Costantini. "Just a Heartbeat Away from One's Body: Interoceptive Sensitivity Predicts Malleability of Body-Representations." *Proceedings of the Royal Society of London B: Biological Sciences* 278.1717 (2011): 2470-76.

Varela, Francisco J. "At the Source of Time: Valence and the Constitutional Dynamics of Affect: The Question, the Background: How Affect Originarily Shapes Time." *Journal of Consciousness Studies* 12.8-10 (2005): 61-81.

Von Economo, Constantin, and Georg N. Koskinas. *Die Cytoarchitektonik der Hirnrinde des Erwachsenen Menschen: Taf.* I-CXII. Atlas. Cornell University: J. Springer, 1925.

Watson, Karli K., Todd K. Jones, and John M. Allman. "Dendritic Architecture of the Von Economo Neurons." *Neuroscience* 141.3 (2006): 1107-12.

Weber, Michel, and Anderson Weekes, eds. *Process Approaches to Consciousness in Psychology, Neuroscience, and Philosophy of Mind.* ,

Albany, NY: State University of New York Press, 2010.

Weber, Michel, and Anderson Weekes. "Sense Perception in Current Process Thought: A Workshop Report." *Mind and Matter* 1.1 (2003): 121-27.

Zelazo, Philip David, and William A. Cunningham. "Executive Function: Mechanisms Underlying Emotion Regulation," In *Handbook of Emotion Regulation,* edited by J. Gross, 135-58. New York: Guilford Press, 2007.

# Contributors

JESSE STERLING BETTINGER

received his Ph.D. in 2015 from the Claremont Graduate University—with a science advisor from Harvey Mudd College—with a dissertation exploring the link between the dynamics of Verlinde's string theoretic description of entropic gravity and Whitehead's actual entities. In the summers he teaches neuroscience at Johns Hopkins University to gifted $10^{th}$–$12^{th}$ graders who test into the Center for Talented Youth. A former collegiate rugby, soccer, and baseball player, he also has ten years of experience coaching soccer at the youth, high school, and college levels. His current research interests include neuro-visceral studies, emotion, interoception, and insular dynamics in the capacity of theories of mind and consciousness. In addition, a chapter of his dissertation develops a method for explaining the conceptual foundations underwriting modern physics without the use of equations.

RONNY DESMET

became a Master of Science in Mathematics in 1983 with a disserta-tion on the mathematics and philosophy of quantum mechanics. After a career in the private sector, he left a position at Sun Microsystems in 2002 to study philosophy. He became a Master of Philosophy in 2005 with a dissertation on the decline of the mechanistic worldview, and a Doctor of Philosophy in 2010 with a dissertation exploring Whitehead's philosophy of mathematics and relativity. Currently, he is an FWO-postdoctoral research fellow (FWO stands for Fund for Scientific Research Flanders), and a member of the Center for Logic and Philosophy of Science at the Vrije Universiteit Brussel (Free University of Brussels). For downloadable versions of most of his publications, see: http://vub.academia.edu/RonaldDesmet/

## TIMOTHY E. EASTMAN

is a consultant in space physics and plasma sciences, based in Silver Spring, Maryland. Concurrently, he is a senior scientist providing support for NASA's Goddard Space Flight Center and has more than 40 years of experience in research and consulting in space physics, space science data systems, space weather, plasma applications, education, and philosophy. He created and maintains a major website for plasma science and applications at plasmas.org. Dr. Eastman has provided key leadership of the nation's research programs in space plasma physics while program manager at NASA Headquarters (1985-1988) and the National Science Foundation (1991-1994). He discovered the Low-Latitude Boundary Layer of Earth's magnetosphere and has published over 100 research papers in space physics and related fields. He has authored more than 30 philosophical papers and reviews, including the editing of four special journal issues on process thought and natural science, in addition to co-editorship of *Physics and Whitehead* (SUNY, 2004, 2009) and *Physics and Speculative Philosophy: Potentiality in Modern Science* (De Gruyter, 2016).

## ARRAN GARE

is Reader in Philosophy and Cultural Inquiry, Swinburne University, Australia, and founder of the Joseph Needham Centre for Complex Processes Research. He has published widely on environmental philosophy, the history of ideas, process metaphysics, the metaphysical foundations of the sciences, philosophy of mathematics, complexity theory, human ecology, the emergent theory of mind, social and cultural theory, Chinese philosophy, ethics, and political philosophy. He is the author of a number of books, including *Postmodernism and the Environmental Crisis* (London: Routledge, 1995) and *Nihilism Inc.: Environmental Destruction and the Metaphysics of Sustainability* (Sydney: Eco-Logical Press, 1996). In 2005 he founded *Cosmos and History: The Journal of Natural and Social Philosophy*, of which he is an editor.

## GARY HERSTEIN

Ph.D., is an independent scholar currently working on issues relating to the work of Alfred North Whitehead, particularly Whitehead's "natural philosophy" from the early 1920s, as this is applied to the philosophy of science, and Whitehead's metaphysics as this relates to the philosophy of logic. His publications include *Whitehead and the Measurement Problem of Cosmology*, Ontos-Verlag (now DeGruyter) 2006; "Davidson on the Impossibility of Psychophysical Laws," *Synthese*, Volume 145, Number 1, May 2005, pp. 45–63; "The Roycean Roots of the Beloved Community," *The Pluralist*, 4.2, Summer 2009; and "Alfred North Whitehead," *The Internet Encyclopedia of Philosophy*, see: http://www.iep.utm.edu/w/whitehed.htm

## FARZAD MAHOOTIAN

is clinical assistant professor in the Liberal Studies program of New York University, and an affiliated scholar with the Consortium for Science Policy and Outcomes at Arizona State University. His interdisciplinary work on the interactions of science and culture is rooted in process philosophy and the chemistry of non-equilibrium systems. His research focus is the impact of metaphor and mythic thinking on the practice of scientific research in the history of science. Recent publications include studies in laboratory ethnography, the philosophy of chemistry, and a co-authored comparison of A. N. Whitehead and C. G. Jung on rational intuition. He received a Ph.D. in Philosophy from Fordham University, a M.S. in Chemistry from Georgetown University, and has worked extensively with NASA to make remote sensing data of the Earth system relevant to science education. His current interests center on the art-science interface where metaphoric imagination and technology activate creativity.

## RONALD P. PHIPPS

is co-of the International Center for Process Philosophy, Science and Education. He was the personal assistant to the former president of the American Philosophical Association, Henry S. Leonard, who was the personal assistant to A. N. Whitehead at Harvard. Under a National Science Foundation Fellowship in theoretical physics Phipps explored the implications of Whitehead's metaphysics to science. He writes on physics, cosmology, education, and economic, social, and ecological development. He served as chair of The US-China Friendship Association and participated in the normalization and development of US-China relations. Phipps is president of the Chamber Players International. He is also a poet and photographer.

## JEAN PAUL VAN BENDEGEM

is at present professor at the Vrije Universiteit Brussel (Free University of Brussels), where he teaches courses in logic and philosophy of science. Van Bendegem studied mathematics and philosophy at the Rijksuniversiteit Gent (University of Ghent) and wrote a doctoral thesis on the problem of strict finitism. While strict finitism is still one of his main research projects, the study of mathematical practice has become equally important, viz., to try to understand what it is mathematicians do when they do mathematics. In addition, he closely follows the discussion about the relations between the sciences and religious worldviews and is interested in possible connections between mathematics and the arts. Van Bendegem is also director of the Center for Logic and Philosophy of Science (www.vub.ac.be/CLWF) and the editor of the logic journal Logique et Analyse (www.vub.ac.be/CLWF/L&A).

## ROBERT J. VALENZA

completed both his undergraduate and graduate education at Columbia University. He came to Claremont McKenna College (CMC) in

1988 to teach mathematics and computer science after two years at Harvey Mudd College. In 1994 he became the W. M. Keck Professor of Mathematics and Computer Science, and a year later he headed a committee to draft a new freshman general education requirement for CMC, one that eventually was transformed into the current Freshman Humanities Seminar (FHS). Reflecting his extensive work with FHS and its predecessor, his position changed, in 2005, to Dengler-Dykema Professor of Mathematics and the Humanities. Dr. Valenza now teaches both FHS and mathematics in equal parts. His research interests are accordingly diverse and include metaphysics, aesthetics, the philosophy of mathematics and science, stylometry, and number-theoretic algorithms.

# Name Index